HANDBOOK OF VALVES

HANDBOOK OF VALVES

Philip A. Schweitzer, P. E.

INDUSTRIAL PRESS INC.
200 Madison Avenue, New York, N.Y. 10016

Library of Congress Cataloging in Publication Data

Schweitzer, Philip A.   1925–
 Handbook of valves.
  1. Valves—handbooks, manuals.

 I. Title.
 TS277.S39   621.8′4   72-5835
 ISBN 0-8311-3026-1

HANDBOOK OF VALVES

Copyright © 1972 by Industrial Press Inc., New York, N.Y. Printed in the United States of America. All rights reserved. This book or parts thereof may not be reproduced in any form without permission of the publishers.

Contents

Preface		ix
1. General Consideration		1
	1.1 Introduction	1
	1.2 Valve Functions	1
	1.3 Valve Selection Factors	2
	1.4 Standards and Specifications	4
	1.5 Abbreviations and Terms	7
2. Valve Design		13
	2.1 Flow Control Elements	13
	2.2 Movement of Flow Control Element	14
	2.3 Sealing Methods	17
	2.4 Methods of Valve Operation	28
	2.5 Valve Connections	36
	2.6 Installation Techniques	39
	2.7 Drain Connections and Bypasses	45
3. Flow Through Valves		46
4. General Purpose Valves		57
	4.1 Gate Valves	58
	4.2 Globe Valves, Angle Valves, Needle Valves	66
	4.3 Y-Type Valves	76
	4.4 Plug Valves	77

CONTENTS

4.5	Ball Valves	87
4.6	Butterfly Valves	93
4.7	Diaphragm Valves	102
4.8	Pinch Valves	112
4.9	Slide Valves	115

5. Check Valves 116

5.1	Swing Check Valves	119
5.2	Tilting Disc Check Valves	122
5.3	Lift Check Valves	125
5.4	Piston Check Valves	129
5.5	Butterfly Check Valves	129
5.6	Spring-Loaded Check Valves	132
5.7	Stop Check Valves	134
5.8	Foot Valves	134

6. Special Service Valves 136

6.1	Pressure Relief Valves	136
6.2	Flush Bottom Tank Valves	151
6.3	Sampling Valves	154
6.4	Solenoid Valves	154
6.5	Pressure Reducing Valves	158
6.6	Backpressure Regulating Valves	162
6.7	Cryogenic Valves	163
6.8	Control Valves	163

7. Valve Packing 164

Appendix 175

Index 177

List of Tables

1–1.	Trim Parts on Common Valves	9
2–1.	Standard Size of Tapping	45
2–2.	Sizes of Bypass Connections	45
3–1.	Resistance to Flow of Fully Opened Valves Expressed in Equivalent Lengths of Clean Schedule-40 Steel Pipe	50
4–1.	Recommended Valve Services	57
4–2.	Size and Operating Ranges of Valves	58
4–3.	Maximum Operating Temperature of Ball Valve Seats	89
4–4.	Body Materials for Diaphragm Valves	104
4–5.	Diaphragm Materials	106
5–1.	Check Valve—Flow Control Valve Combinations	117
5–2.	Sizes and Operating Ranges of Check Valves	117
6–1.	Constant C for Gas or Vapor Related to Specific Heats $(K = C_p/C_v)$	141
6–2.	Classification of Hazardous Locations	157
7–1.	Packing Recommendations	169
7–2.	Packing Description	171

Preface

Valves are so commonplace and so widely used that relatively little attention is paid to them by many maintenance, operating, and construction personnel; engineers and specifiers — until the valves fail or do not perform exactly as required. In most cases of valve failure the fault can be traced to the misapplication of the valve: in other words, the wrong valve was used for the service.

For many years the selection of valves was a relatively simple matter since the varieties to choose from were somewhat limited. Technological advances made in the development of new materials of construction and more stringent requirements imposed on piping systems (and valves) have led to many new developments in the valve industry. Modifications have been made in the older conventional types (such as gates, globes, plugs, etc.) and new designs of valves have also emerged.

At times it can be very bewildering to look through various valve catalogs and uncover perhaps a half dozen different designs of valves, all of which appear to be suitable for a particular application. The logical question often asked in such a situation is: "Which valve shall I use — which valve is best suited for the application?"

It is the purpose of this book to help the valve user and/or specifier to make a logical and correct choice. Valve selection starts with an understanding of the function the valve is expected to perform and the factors which affect its performance, such as the properties of fluids going through the valves, fluid friction losses, operating conditions, materials of construction, and size, all of which are discussed and explained in the text.

The design of each type of valve, with its possible variations, is taken up, and the advantages and disadvantages of each are explained so that the proper valve can be matched to perform the function required, taking into consideration the requirements of the system in which it is to operate. A large number of line and halftone illustrations are furnished to aid the reader in quickly grasping the similarities and differences of the various designs.

Also included is a chapter which describes how to determine the proper valve size and the resultant pressure drop. The pressure drop may be determined in terms of equivalent pipe lengths and/or in other units of pressure measurement.

Whether a valve is to be operated manually or by means of an automatic control device has no bearing on the type of valve selected for the application. Details of various means of operating valves manually are covered in the book. There is also a section dealing with solenoid valves and an explanation of air, hydraulic, and electric motor operators for valves. However, the design of automatic control systems for valves is not covered. That is a complete subject by itself.

Related practical aspects of valving, such as valve location, valve care (before and after installation), installation, operation, and maintenance are discussed and guides set forth.

Another chapter deals with the important subject of valve stem packing and includes packing recommendations for specific types of services.

Anyone involved in the selection, installation, operation and/or maintenance of valves — regardless of the industry or application — should find this book helpful.

1

General Consideration

1.1 INTRODUCTION

A substantial portion, approximately 5 percent, of the total capital expenditures for the chemical process industry is used for the purchase of valves. In terms of numbers of units, valves are exceeded only by piping components.

With these facts in mind it is readily seen why care should be taken and thought given to the selection of the valves to be used in a particular installation.

It is the purpose of this book to provide a source of reference where the specifying engineer, maintenance engineer, and/or anyone else involved in the valve selection procedure can look to find guides for the appropriate selection, installation, and maintenance of valves of all types.

1.2 VALVE FUNCTIONS

Valves are used in piping systems and on processing vessels or tanks for a variety of reasons. The first step in valve selection is to determine exactly what is expected of the valve — that is, what function the valve is to perform after it has been installed. Proper evaluation of this function will, more than any other single factor, narrow down the types of valves suitable for the application. General valve functions can be defined as follows:

1. On-off service
2. Throttling service

3. Prevention of reverse flow
4. Pressure control
5. Special functions
    A. Directional flow control
    B. Sampling service
    C. Limiting flows
    D. Sealing vessel and/or tank outlets nozzles
    E. Other miscellaneous functions.

Appendix I on page 175 tabulates the generally recommended functions for which the various types of valves are most commonly applied. However, there are special designs of the various types which permit them to operate in other categories. These special designs are discussed under the individual valve headings.

1.3 VALVE SELECTION FACTORS

Usually more than one type of valve is suitable to perform a specific function. In order to narrow the selection down to the one most applicable it is necessary to investigate the factors which affect the individual valve's performance and the effect that a particular valve has on the materials being handled.

Fluid Properties

The properties of the fluid being handled must be known. These properties include its specific gravity, viscosity, corrosiveness, and abrasiveness. Fluid, as used here, is a general term which can mean gas, vapor, or slurry, as well as a pure liquid material.

An analysis of the system should be made to determine whether or not more than one fluid will pass through the valve. Special care should be taken in examining what materials (gaseous and liquid) could possibly come into contact with a valve connected to the head of a processing vessel. A valve on a feed line to a processing vessel will naturally be selected based on the material flowing through the line. However, other factors must be taken into account. Once the vessel has been charged and placed into operation it is very possible that entirely different fluids will contact the vessel side of the valve as a result of the reaction taking place in the vessel.

Fluid Friction Losses

The various types of valves exhibit pressure drops to varying degrees. A system requirement of limited available pressure drop can often influence a valve selection. This factor must be considered in valve selection.

A typical system requiring limited pressure drop is the suction piping for a pump. When designing such a system the Net Positive Suction Head (NPSH) must be taken into account. NPSH combines all of the factors limiting the suction side of a pump: internal pump losses, static suction lift, friction losses, vapor pressure, and atmospheric conditions. It is necessary to differentiate between required NPSH and available NPSH.

Required NPSH refers to internal pump losses and is determined by laboratory test. Available NPSH is a characteristic of the suction system, and can be calculated. By definition it is the net positive suction head, above the vapor pressure, available at the suction flange of the pump to maintain a liquid state. Since there are also internal pump losses (required NPSH), the available NPSH in a system must exceed the pump required NPSH.

Available NPSH may be calculated using the following formula:

$$\text{NPSH}_{(a)} = H_P \pm H_Z - H_F - H_{VP}$$

Where:

$\text{NPSH}_{(a)}$ = Available NPSH expressed in feet of liquid

H_P = Absolute pressure on the surface of the liquid where the pump takes suction, expressed in feet. This could be atmospheric pressure or vessel pressure (pressurized tank)

H_Z = Static elevation of the liquid above or below the center line of the pump impeller, expressed in feet

H_F = Friction and velocity head loss in the piping, also expressed in feet

H_{VP} = Absolute vapor pressure of the fluid at the pumping temperature, expressed in feet of liquid.

Pressure drop through a valve or valves on the suction piping becomes part of the H_F term. It is advantageous to keep this term as small as possible; therefore the pressure drops through any valves in the line become critical.

Operating Conditions

Maximum and minimum coincident pressures and temperatures must be known. A valve choice, particularly in a corrosion-resistant material of construction, can be influenced greatly by these factors. This is especially true when plastic or lined valves are under consideration.

Establishment of the actual operation condition of each valve will simplify the valve selection procedure.

Materials of Construction

This factor is directly related to the fluid properties of corrosiveness and abrasiveness. In dealing with extremely corrosive or abrasive materials the choice of valve may be limited by availability of valves in suitable materials of construction.

At times it is desirable to consider the material of construction of the body separately from the material of construction of the trim (stem, seat ring, disc), in order to make the most economical selection. For certain types of lined valves, such as diaphragm valves, the lining material will normally be different from the elastomeric diaphragm.

The combination of operating pressure and operating temperature will influence the design criteria of the valve. The design temperature and pressure, along with the fluid characteristics, will determine the allowable materials of construction.

Valve Size

Because all types of valves are not available in a complete range of sizes, it is necessary to know what valve sizes will be required to perform each function. In addition, the economy of one type of valve over another can change as the valve size changes. The problem of available valve sizes is also prevalent in certain corrosion-resistant construction materials.

1.4 STANDARDS AND SPECIFICATIONS

Various technical societies, trade associations, and governmental agencies have established valve standards and/or specifications. Some of the standards and specifications are recognized industry-wide by valve manufacturers, some are applicable only for specific valve services or uses, while others are applicable to all valves used within a specific industry.

Most of these standards and/or specifications are designated by letters — the most common of which are listed below:

USASI (USAS) — United States of America Standards Institute — establishes certain basic dimensions on valves, fittings, and threads.

ASTM — American Society for Testing Materials — establishes and writes chemical and physical requirements of all materials used in the manufacture of valves and fittings.

API — American Petroleum Institute — establishes purchasing standards on valves and fittings for the Petrochemical industry.

AWWA — American Water Works Association — establishes standards on iron gate valves to be used in a recognized water supply system.

MSS — Manufacturers Standardization Society of the Valves and Fittings Industry — maintains standards on dimensions, marking, boss locations for drains and by-passes, testing, and other similar type standards.

UL — Underwriters Laboratories — establishes design and performance standards on valves and fittings used in fire protection service and handling of hazardous liquids.

FM — Associated Factory Mutual — establishes standards similar to Underwriters', but is employed by Mutual Fire Insurance Companies.

AAR — Association of American Railroads — establishes design and dimensional standards on bronze valves and 300-pound malleable pipe fittings for use by railroads.

Federal Specs — Federal Government Specification Standards — established by U.S. agencies for design, dimensions, materials, and tests on valves, fittings, and unions, such as WW-V-51d on bronze globe, angle, and check valves.

Military Specs ("MilSpecs") — U.S. Military Specifications and Standards — established for design, dimensions, materials, and tests on items for use by the Armed Forces. Valves are covered when purchases are required.

USCG or Marine Regulations and Military Specs — U.S. Coast Guard Standards — in general cover materials and testing requirements on valves and fittings for use aboard ship.

NFPA — National Fire Protection Association — establishes design and performance standards on valves and fittings used in fire protection service.

ASME — American Society of Mechanical Engineers — establishes codes covering pressure-temperature ratings, minimum wall thicknesses, metal ratings, thread specifications, etc., for valves made of materials meeting ASTM specifications.

Metal and Alloy Designations

As just explained, the ASTM establishes physical requirements and chemical composition of materials which are used in the manufacture of the various valve components. Each of these materials is given an ASTM specification number. When valve construction is being compared between manufacturers, reference to the ASTM specification numbers is helpful. If the manufacturers supply the ASTM specification covering their valves (and most manufacturers do), a quick comparison and analysis of the valve material can be made. Listed below are the more common materials used in valve manufacture and the corresponding ASTM designations. You will note that in several cases a single material may be covered by more than one ASTM specification. This is because of differing chemical compositions and/or differing physical properties available under the particular generically named material. For example, two silicon brasses are shown. A comparison of their properties would be as follows:

		ASTM B-198 Grade 13B	ASTM B-371 Alloy A
Chemical Properties %	Copper Lead Tin Zinc	Remainder Max. 0.50 3.00–5.00 12.00–16.00	80.00–83.00 Max. 0.30 3.50–4.50 Remainder
Mechanical Properties	Tensile Strength, psi Yield Strength, psi Elongation Min. % in 2"	60,000 24,000 16	80,000 (to 1") 40,000 15

Metals	ASTM Designations
Bronzes and Brasses	
High Tensile Steam Bronze	B-61
Steam Bronze	B-62
Cast Silicon (Everdur 1000)	B-198 Grade 12A
Silicon Brass	B-198 Grade 13B
Silicon Brass	B-371 Alloy A
Wrought Silicon (Everdur 1012)	B-98 Alloy D
88-10-2 Bronze	B-143 Class 1A
Ampco, Grade C-3	B-148 Alloy 9C
Ampcoloy, Grade B-2	B-148 Alloy 9B
Brass Rod	B-16
Naval Brass	B-21 Alloy A
Brass Tubing	B-135 Alloy G
Phosphor Bronze	B-134 Alloy B-2
Bronze Rod	B-140 Alloy B
Irons	
Cast Iron	A-126 Class A
Cast Iron	A-126 Class B
High Tensile Cast Iron	A-126 Class C

Malleable Cast Iron	A-47 Grade 35018
Malleable Cast Iron	A-47 Grade 32510
Ni-resist Gray Cast Iron	A-436 Type II

Cast Steels

Carbon Steel, Cast	A-216 Grade WCB
0.15% Moly. Steel, Cast	A-217 Grade WC1
Cr. Moly. Steel, Cast	A-217 Grade WC-6
Cr. Moly. Steel, Cast	A-217 Grade WC-9
4.6% Cr. Moly. Steel	A-217 Grade C5
8–10% Cr. Moly. Steel	A-217 Grade C-12
Carbon Steel, Cast	A-352 Grade LCB
Carbon Moly. Steel, Cast	A-352 Grade LC-1
3.5% Nickel Steel, Cast	A-352 Grade LC-3

Stainless Steels

18.8 S Cast	A-351 Grade CF8 (Type 304)
Wrought	A-276 Type 304
Wrought	A-276 Type 303
18-8S Mo, Cast	A-351 Grade CF-8M (Type 316)
Wrought	A-276 Type 316
18-8S Cb, Cast	A-351 Grade CF-8C (Type 347)
Wrought	A-276 Type 347
11.5–13.5 Cr. Steel	A-182 Grade F-6
11.5–13.5 Cr. Steel	A-276 Type 416
Heat-Resisting 25–12 Cast	A-297 Grade MH
Heat-Resisting 25–20 Cast	A-297 Grade MK
Heat-Resisting 25–20, Wrought	A-182 Type F 310
Heat-Resisting 15–35	A-297 Grade HT

Nickel Alloys

Nickel Cast	A-296 CZ100
Nickel Wrought	B-160
Monel Cast	A-296 M-35W
Monel Wrought	B-164 Class A
Hastelloy "B" Cast	A-296 N-12M
Hastelloy "C" Cast	A-296m CW-12M

Aluminum

No. 356 T6 Cast	B-26 Grade SG70AT6

1.5 ABBREVIATIONS AND TERMS

There are many standard abbreviations used to describe types of valves, features of valves, and the different valve parts. These abbreviations are used constantly and consistently by valve manufacturers and should be understood. The more common abbreviations with their meanings are listed below:

OS & Y	Outside Screw and Yoke — describes construction of a valve
NRS	Non-Rising Stem — describes construction of a valve and operation of the valve stem
RS	Rising Stem — describes construction of a valve and operation of the valve stem
WOG	Water, Oil, Gas — pressure rating applying to relatively cool liquids and gases
CWP	Cold Working Pressure
WSP or SP	Allowable Working Steam Pressure
"LPG"	Liquified Petroleum Gas
IBBM	Iron Body — Bronze Mounted — describes valve construction
All Iron	All Iron construction
TE or SE	Threaded End connection
FE	Flanged End connection
BWE or WE	Butt Welding End connection
SWE	Socket Welding End connection
SJ	Solder Joint End connection
SB	Silver Braze End connection
SIB	Screwed Bonnet
UB	Union Bonnet
BB	Bolted Bonnet
ISRS	Inside Screw Rising Stem
ISNRS	Inside Screw Non-Rising Stem
SW	Solid Wedge
DW	Double Wedge
DD	Double Disc
TD	TFE Disc (Teflon,[1] Halon,[2] etc.)
FF	Flat Face Flange
RF	Raised Face Flange
LFM	Large Male and Female Flange
SMF	Small Male and Female Flange
LF	Large Female Flange
SF	Small Female Flange
LM	Large Male Flange

[1] Trademark E. I. duPont de Nemours & Co., Inc.
[2] Trademark Allied Chemical Corp.

SM	Small Male Flange
LTG	Large Tongue and Groove Flange
STG	Small Tongue and Groove Flange
LT	Large Tongue Flange
ST	Small Tongue Flange
LG	Large Groove Flange
SG	Small Groove Flange
RTJ	Ring-Type Joint Flange
Int S	Integral Seat
Ren. S	Renewable Seat
IPS	Iron Pipe Size
PSI	Pounds per Square Inch.

These abbreviations will be used throughout the book.

Terms

There are many unique terms in valve design, operation, and performance which are used in valve specifications. These terms are defined and explained in this section.

Trim designates those parts of a valve which are replaceable and which usually take the most wear. The common trim parts are shown in Table 1-1 for various types of valves.

Table 1-1. Trim Parts on Common Valves

Type Valve	Gate	Globe	Swing Check	Lift Check
Trim parts	Stem Seat ring Wedge Pack under pressure bushing on steel valves	Stem Seat ring Disc Pack under pressure bushing on steel valves Disc nut	Disc Disc holder Disc carrier Disc nut Side plug Carrier pin or holder pin Disc nut pin Seat rings	Disc Disc guide Seat rings

Straight-Through Flow refers to a gate valve (or other type of valve) in which the wedge (or closure element) is retracted entirely clear of the waterway so that there is no restriction to the flow.

Full Flow refers to relative flow capacity of various valves.

Throttled Flow refers to the suitability of a valve for throttling service; that is, for service when the valve is not fully open, or, more particularly, when the valve is nearly closed. In very close throttling there is danger of wire drawing (see page 10). Gate valves are not recommended for throttling service since the accompanying turbulence will cause the wedge to vibrate against the seat with resulting damage.

Wire-Drawing means the premature erosion of the valve seat caused by excessive velocity between the seat and seat disc. An erosion pattern is left as if a wire had been drawn between the seat surfaces. Excessive velocity can occur when a valve is not closed tightly. A WOG disc is the best defense against wire-drawing because its resiliency makes it easier to close tightly. Steam discs are harder materials and must be closed more carefully to prevent wire-drawing. Even closed tightly, a valve in steam service may start leaking later because of uneven cooling of its parts. These valves should be checked periodically until fully cooled as a precaution against leaking. Another protective measure against leaking is to install a globe valve with pressure over the seat *whenever conditions permit*. The pressure will act to maintain the closure of the valve if uneven cooling should occur.

Dirty or Abrasive Flow refers to flow conditions and the ability of a valve to function properly and remain usable when the fluid flowing through it contains solids or abrasives in suspension or entrainment. WOG disc material is least damaged by abrasives. Abrasive flow accelerates the wire-drawing effect.

Noncritical Applications are those applications in domestic water services, domestic heating systems, plant utilities and other systems where pipe or valve failure would not seriously endanger persons or property or cause economic losses.

Critical Applications are a wide range of special or unusual applications where pipe or valve failure would seriously endanger persons and property or cause economic loss. Appropriate pressure ratings and materials of construction should be used to satisfy the critical need of the application.

Frequent Operation refers to the cycle of operation of a valve. Globe valves may be operated thousands of times without severe wear because the seating action primarily involves compression of the disc against the seat. Conventional gate valves close with a sliding action that causes wear on the seat, especially under abrasive conditions. Consequently, the life of a globe valve will be longer in the same service than a gate valve when there is frequent operation.

Replaceable Disc refers to the disc of any valve which can be replaced. In a gate valve the seats in the body undergo wear comparable to the wedge disc. Replacing a worn wedge disc will not, in most cases, result in a usable valve. Replacing the metallic disc in a globe valve may not result in a usable valve either, if the seat is damaged.

Replaceable Seat refers to any valve in which the seat can be replaced. There are many valve designs having disc and body seat rings made of materials different from the valve body. These seating parts are used to gain greater wear resistance and extended life, not necessarily because of the "renewability" feature. However, the seats can be replaced when worn.

Metal-to-Metal Seats refer to a valve having a metal disc and metal seat. Valve constructions are available which utilize various materials for discs and/or seats other than metal. Included are materials such as buna-N, TFE, Kel-F, etc. These composition disc materials tend to provide more positive closure for the valve than a metal-to-metal seated valve. Consequently, the metal-to-metal seat construction should only be specified when the fluid involved will damage the composition disc and/or when the operating conditions of temperature and/or pressure exceed the limitations of the composition material; or when the flow demands a gate valve.

Visual Indication of Valve Open or Closed is the degree of closure that a rising stem shows in gate or globe valves, or the position the handle indicates on plug cocks or ball valves.

On gate or globe valves a small amount or no stem showing between the packing nut and handwheel indicates the valve is closed or very nearly closed. Considerable stem showing indicates the valve is more or fully open. This is a relative measure and varies with valves being observed. Some valves are equipped with an indicating device such as a pointer which rides up the stem and follows a scale graduated in percentage opening of the valve. Nonrising stem valves do not provide this indication.

Handles on plug cocks and ball valves are usually mounted so that the handle when at 90 degrees to the pipelines indicates the valve is closed. When the handle is oriented along the axis of the pipeline, the valve is full open.

Limited Installation Space refers to the space required for valve operation after installation. Check valves operate internally and therefore require no external operating space and can be installed in limited areas. Nonrising stem valves require little more than the space needed for installation. The handwheels rotate without rising. Rising stem valves require extra space after installation to accommodate the rising of the handwheel as the valve is opened fully.

Plug cocks and ball valves require space to permit operation of the handle across the pipeline.

Installation Position is the position of the valve stem when installed. Most valves are not limited as to the position of installation. However, all nonspring loaded check valves do depend upon gravity action to close when flow stops. Care in positioning should be taken in these cases.

TFE is the designation of polytetrafluoro-ethylene resin. This base resin is manufactured under the trade names of Teflon-TFE by E. I. duPont de Nemours & Co., Inc. and Halon-TFE by the Allied Chemical Corp. There are also other manufacturers. The chemical resistance of TFE is unique in comparison to other materials. It is virtually inert, chemically, to all commercially available chemicals and solvents with the exception of molten alkali metals, fluorine, and chlorine trifluoride at high temperatures and pressures. This inertness exists throughout the entire allowable working temperature range.

2

Valve Design

The details of valve design can be divided into three individual design problems. Each phase or problem can be treated separately. Of primary importance to the proper operation of the valve is the design of the flow-control element. This is that portion of the valve, with its various components, which actually controls the flow of fluids through the valve.

Second, in logical sequence, is the design of the mechanism whereby the flow control element may be adjusted to permit control of the rate of flow through the valve.

The third area of consideration is the isolation of the mechanism governing movement of the flow control element from the fluid being handled in the valve (sealing methods).

When valve design is said to consist of the three areas just mentioned, it is understood that only the design of the operational portions of the valve are being considered. Other design problems can and certainly do exist — such as mechanical strength, dimensional arrangements, types of end connection, etc. However, this book is only concerned with the design of the operational portions of the valves.

2.1 FLOW CONTROL ELEMENTS

Ancient Pompeians used tapered plug valves to control the flow of fluids in closed pipelines as early as 79 A.D. This concept, although improved upon, is still used today along with several other flow-control means. From the multiplicity of valves presently available, one would tend to assume that there are also a multiplicity of ways in which the flow through a closed pipe can be controlled. Actually, there are only four basic means of controlling flow in a pipe, and one of these is the method used by the ancient Pompeians.

The four basic methods employed to control flow through a valve are:

1. Move a disc or plug into or against an orifice such as is done in the globe, angle, Y, and needle valves.
2. Slide a flat, cylindrical, or spherical surface across an orifice such as is done in the gate, plug, ball, slide, and piston valves.
3. Rotate a disc or ellipse about a shaft extending across the diameter of a circular casing as is done in butterfly valves and dampers.
4. Move a flexible material into the flow passage such as is done in diaphragm and pinch valves.

All valves, presently available, control flow by one or more of the above methods. Many refinements have been made and improvements incorporated into the designs as technology has improved and newer, more advanced materials have become available. Each type of valve design has its place in today's industry. Each type of valve has been designed for specific functions; and when used to meet these functions, the valve will give good service and have a long life.

Specific details of each type of construction will be discussed under the individual type of valves employing such construction. Although only four basic means of controlling flow are listed, it will be found that many modifications exist within each major type.

2.2 MOVEMENT OF FLOW CONTROL ELEMENT

This phase of valve design is actually divided into two sections — section one being the mechanism which imparts movement to the flow control element, and section two being the means whereby this mechanism is operated. The latter portion is referred to as *The Methods of Valve Operation* and will be covered in Section 2.4, while the former is covered in this section.

Movement of the flow control element is accomplished by means of a stem which is fixed to the flow control element and rotates, moves endwise, or combines both movements in order to set the position of the flow control element. In most types of valve the stem extends to the outside of the valve. Exceptions to this are check valves, and some safety and regulating valves which are operated by the force of the fluid within the pressure zone.

Rotating Stems

This type of movement will be found on nonrising-stem gate valves, rotating-disc gate valves, ball valves (usually quick opening), butterfly valves, and most plug valves and cocks.

Endwise Stem Movement Without Rotation

This type of movement will be found on outside-screw and yoke gate valves (OS & Y); quick opening gate valves; globe and diaphragm valves; slide, piston and sleeve valves; and outside-spring safety and relief valves. On the OS & Y gate valves only the handwheel rotates with the valve stem raising through the handwheel. This is a somewhat awkward valve to operate in smaller sizes since the hand cannot be placed across the handwheel while it is turning.

Stem Rotates and Moves Endwise

Globe, angle, Y, and needle valves; rising-stem gate valves; lift-type plug valves; and most diaphragm and pinch valves employ this type of movement.

The majority of valves employ a threaded stem for the movement of the flow control element. Notable exceptions are butterfly valves, ball valves, solenoid valves, all operated valves, and safety relief valves, regulating valves, and check valves. When selecting valves with threaded stems, several factors must be considered, such as:

1. Corrosiveness of the fluid being handled
2. Corrosiveness of the surrounding atmosphere
3. Operating temperature
4. Available headroom
5. Desirability of having the stem position indicate the amount of valve opening.

Globe valves are available with either an inside or outside screw rising stem, which both rotates and moves endwise providing visual indication of the degree of valve opening.

Gate valves are available with an inside screw nonrising stem (Fig. 2-1) and an inside screw rising stem (Fig. 2-2). The outside-screw gate valve has a threaded stem which moves endwise only.

It can be seen from Fig. 2-1 and Fig. 2-2 that the stem threads on inside-screw valves are exposed to the process fluid being handled. Because of this feature this type of valve operation should not be used on lines handling corrosive materials, slurries, or where the operating temperature is elevated.

Rising-stem gate or globe valves of this design provide visual indication of the degree of valve opening, but adequate headroom must be provided to allow for the rise of the stem to the fully opened position. A nonrising-stem gate valve is ideal for use in areas having limited head

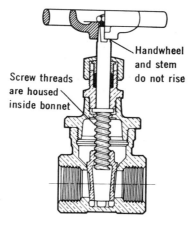

Fig. 2-1. Inside screw nonrising stem (ISNRS) gate valve.

room. Since only the spindle turns, stem wear is minimized. The wedge rises on the threaded portion of the stem.

In outside-screw and yoke (OS & Y) valves — refer to Fig. 2-3 — the spindle packing is between the threaded portion of the stem and the process fluid, preventing contact with the threads by the process fluid. Consequently this type of valve is recommended for use on systems handling corrosive media, slurries, and systems operating at elevated temperatures. The position of the stem provides a visual indication of the degree of valve opening. In addition, the external screw also permits

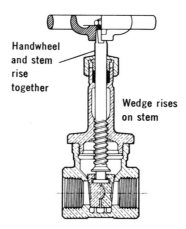

Fig. 2-2. Inside screw rising stem (ISRS) gate valve.

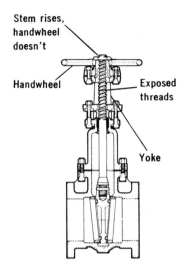

Courtesy of NIBCO Inc.

Fig. 2-3. Outside screw and yoke (OS&Y) gate valve.

easy lubrication of the thread. Care should be taken to protect the exposed threads from damage and headroom must be provided to allow for the rise of the stem when the valve is in the fully opened position.

Outside screws present a disadvantage when the external atmosphere contains corrosive fumes or vapors. Since the threads can be corroded, valve operation becomes difficult and at times impossible. In such cases an inside screw should be employed, providing the media being handled is not corrosive. There are many situations where both the external atmosphere and the internal media are corrosive. In such cases special bonnet seals must be employed. These are discussed in Section 2.3.

Ball valves, plug valves, and butterfly valves usually control the flow by means of a rotating stem without the use of threads. As the stem is rotated, the flow-control element is similarly rotated providing a larger or smaller passage for the process fluid.

Gate and globe valves are also available with sliding stems where quick opening or quick closing is required.

Regardless of type of valve, if quick closing is a feature of operation, provision must be made in the piping system to compensate for hydraulic shock.

2.3 SEALING METHODS

There are four places in a valve where sealing is required. One, of course, is the prevention of leakage of the process fluid downstream when the valve is in the closed position. The remaining three are concerned with the leakage of the process fluid to the outside and/or the prevention of leakage of air into the system when the line is operating under vacuum. These latter seals must be made at the stem, the valve end connection, and where the bonnet joins the valve body. Because of the movement involved, stem sealing is more difficult to accomplish than the other two.

Since the valve end connections are actually external of the valve design itself, these seals will be considered in the section on valve connections.

Flow Seals

In order to provide an adequate seal against the flow of the process fluid when the valve is in the closed position, a tight closure must be provided between the flow control element and the valve seat. These components must be designed so that pressure and/or temperature changes, as well as strains caused by connected piping, will not distort or misalign the sealing surfaces.

In general, three types of seals are employed: a metal-to-metal contact; metal in contact with a resilient material; and metal in contact with metal containing a resilient material insert in its surface. With the advent of plastics, valves have become available in a variety of plastic formulations. The generalization of the three types of seals is still valid in plastic valves if plastic is substituted for metal in the above categories. The same analogy is applicable in valves having lined interiors — glass-lined, TFE-lined, rubber-lined, etc.

The greatest strength is obtained from a metal-to-metal seal, but metal seizing and galling may occur. A resilient seal is obtained by pressing a metal surface against a rubber or plastic surface. This type of seal provides a tighter shutoff and is highly recommended for fluids containing solid particles, although in general it is limited to less severe usage or to services where the pressure is not high. Solid particles, which may become trapped between the sealing surfaces, are forced into the soft surface and thereby do not interfere with the closure of the valve. A metal in contact with metal containing a resilient material inserted into its surface provides a primary resilient seal and a secondary metal-to-metal seal. This type of seal can be used at relatively high pressures.

Stem Seals

The most common method of stem sealing is the use of a stuffing box containing a flexible packing material such as graphite-asbestos, TFE-asbestos, TFE, etc. TFE is of particular importance for corrosive applications. The packing may be solid, braided, or a loose fill of granulated

TFE, asbestos fibers and TFE, as well as other compositions. Packing designs include square, wedge, and chevron rings. Occasionally sealing is accomplished by means of "O" rings. A discussion of various valve packings will be found in Section 6.

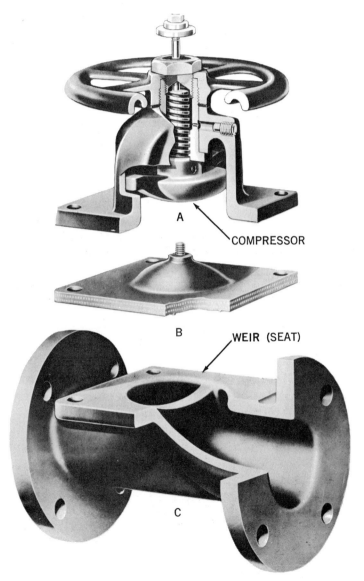

Courtesy of Hills-McCanna Div. Pennwalt Corp.

Fig. 2-4. Saunders patent diaphragm valve. A. Bonnet assembly; B. diaphragm; C. body.

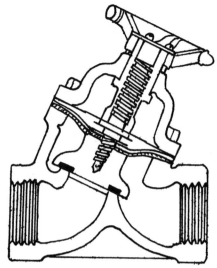

Fig. 2-5. Globe-type valve with diaphragm bonnet seal.

In order to retain the pressure of the fluids inside of the valve it is necessary to compress the packing. This is accomplished by means of a packing nut or gland which compresses the packing into the stuffing box and against the stem. On occasion a spacer or lantern gland is placed in the stuffing box to separate the packing into an upper and lower section.

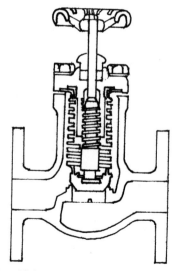

Fig. 2-6. Globe-type valve with bellows bonnet seal.

This permits the introduction of lubricant or inert sealant into the stuffing box through the side. Such a sealing technique requires periodic inspection and maintenance. It is occasionally necessary to take-up (tighten) the stuffing-box nut to maintain adequate packing compression to prevent leakage. Invariably if a valve has not been operated for a period of time, the stuffing box nut must be taken-up when the valve is operated since leakage will occur.

A conventional stem and stuffing box arrangement is unsatisfactory when there must be absolutely no leakage to the outside. Such would be the case when corrosive or extremely hazardous materials are being handled. For such applications a number of valves using packless methods are available.

One such group of valves employs an elastomeric diaphragm between the bonnet and the body. In most cases this diaphragm is pushed down into the flow path by a compressor component which is attached to the stem and thereby also acts as the flow control element. Such a valve is shown in Fig. 2-4.

A globe-type valve, shown in Fig. 2-5, is available with a diaphragm that isolates the working parts from the process fluid.

Another type of packless valve employs a metallic bellows, rather than a flexible diaphragm. This construction is shown in Fig. 2-6. These valves are especially good for operation under very high vacuum. A

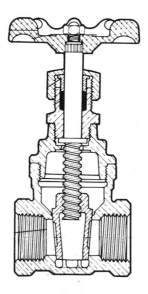

Courtesy of NIBCO Inc.

Fig. 2-7. Globe valve with screw-in bonnet.

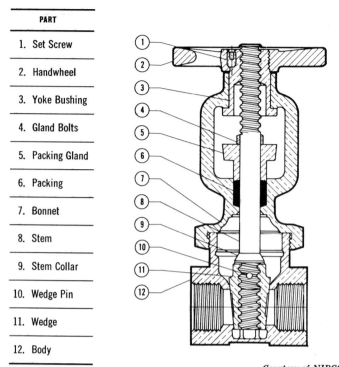

PART
1. Set Screw
2. Handwheel
3. Yoke Bushing
4. Gland Bolts
5. Packing Gland
6. Packing
7. Bonnet
8. Stem
9. Stem Collar
10. Wedge Pin
11. Wedge
12. Body

Courtesy of NIBCO Inc.

Fig. 2-8. Globe valve with screw-on bonnet.

stuffing box is usually provided above the bellows to prevent leakage in the event of bellows failure.

Another design to prevent leakage to the atmosphere employs the use of a double stuffing box with a lantern ring in the middle and a piped drain.

Bonnet Seals

The bonnet is that component of the valve which provides a closure for the body. Normally it is necessary to remove the bonnet in order to gain access to the valve seat and flow control element for purposes of repair and/or replacement. There are three main types of bonnets — screwed, union, and flanged (bolted) as well as several designs for specific applications.

Screwed Bonnet

There are two types of screwed bonnets — the screw-in bonnet (Fig. 2-7) and the screw-on bonnet (Fig. 2-8). These are the simplest and

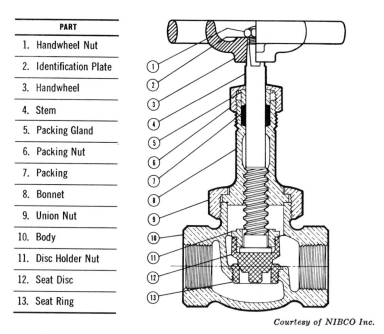

PART
1. Handwheel Nut
2. Identification Plate
3. Handwheel
4. Stem
5. Packing Gland
6. Packing Nut
7. Packing
8. Bonnet
9. Union Nut
10. Body
11. Disc Holder Nut
12. Seat Disc
13. Seat Ring

Courtesy of NIBCO Inc.

Fig. 2-9. Globe-type valve with union bonnet.

most economical types of bonnet designs and are usually found on the smaller valves and limited to low-pressure services.

The operation of screwing the bonnet firmly to the body neck to make a tight seal tends to distort the neck of the bonnet. Consequently it is difficult to make a tight seal again after the valve has been taken apart for maintenance purposes. In addition, it is possible to loosen this seal during normal valve operation or to accidently unscrew the bonnet. Therefore this type of construction should only be used where dismantling will be infrequent and where a minimum of shock and vibration will be encountered.

Union Bonnet

This style of bonnet, shown in Fig. 2-9, provides a quick, easy method of coupling and uncoupling the bonnet from the valve body. The bonnet is provided with a loose nut or union bonnet ring over the bonnet which is screwed onto the valve body while the bonnet is held securely in place. Since all parts are in compression and held firmly in place, distortion is unlikely and the bonnet can be detached and tightly resealed in place any number of times. Union bonnets are usually found on the smaller valves and are recommended for use where frequent dismantling of the

PART
1. Handwheel Nut
2. Handwheel
3. Stem
4. Bonnet
5. Gland Nuts
6. Gland Studs
7. Gland
8. Packing
9. Hex Head Capscrew
10. Body Gasket
11. Disc Holder Nut
12. Disc Holder
13. Disc
14. Disc Plate
15. Seat Disc Nut
16. Seat Ring
17. Body

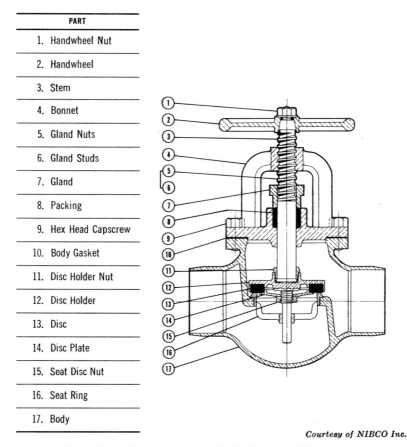

Courtesy of NIBCO Inc.

Fig. 2-10. Globe-type valve with flat-face bolted bonnet.

valve is necessary for maintenance purposes. Because the union nut is separate from the bonnet, there is no danger of loosening the bonnet assembly accidentally while operating the valve.

Bolted (Flanged) Bonnet

This type of bonnet connection is generally used in larger valves and in valves where corrosive media are being handled and/or where high temperatures or pressures will be encountered. Like a flanged piping joint, the bonnet flange is tightened to a similar body flange using a suitable gasket between the faces. The flange joint can be of the flat-faced, male and female, or tongue and groove designs. A flat-faced bolted bonnet is shown in Fig. 2-10. Figure 2-11 shows the male and female joint.

Fig. 2-11. Male and female face bolted bonnet.

In the smaller valves, particularly chemical service valves where frequent disassembly is required for maintenance, the bonnet is often secured by a U-bolt that passes around the body as shown in Fig. 2-12.

These types of bonnets provide a strong connection yet can be easily

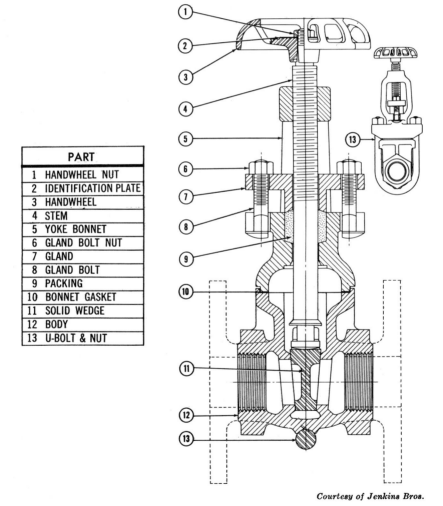

PART	
1	HANDWHEEL NUT
2	IDENTIFICATION PLATE
3	HANDWHEEL
4	STEM
5	YOKE BONNET
6	GLAND BOLT NUT
7	GLAND
8	GLAND BOLT
9	PACKING
10	BONNET GASKET
11	SOLID WEDGE
12	BODY
13	U-BOLT & NUT

Courtesy of Jenkins Bros.

Fig. 2-12. U-bolt bonnet valve.

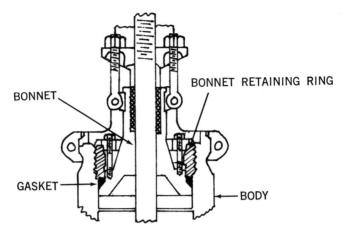

Fig. 2-13. Pressure-sealed bonnet.

disassembled and reassembled without causing any damage to the valve. Whenever the valve is disassembled it is good practice to use a new gasket on reassembly. It can be difficult to obtain a leak-tight joint when reusing the original gasket.

Pressure-Sealed Bonnet

This is one of several types of bonnet seals available for high-pressure and high-temperature services. These assemblies offer a more compact and lightweight design than conventional flanged bonnets.

The pressure sealed bonnet, as shown in Fig. 2-13, uses the fluid pressure in the line to effect a seal. This technique involves fitting the bonnet into the body, followed by a ring gasket or ring seal that is retained within the body either by a segmented ring fitted into a groove in the body or by a retaining ring screwed into the body. The initial seal is obtained by lifting the bonnet upward against the gasket or seal ring by means of bolts. When in service, the internal fluid pressure acts on the underside of the bonnet, forcing it against the gasket or seal ring. As the fluid pressure increases, the tightness of the seal is increased. The bonnet can be removed for maintenance purposes.

Lip-Sealed Bonnet

This style is also used where high pressures and/or high temperatures will be encountered. As shown in Fig. 2-14, the bonnet is screwed into the body until contact is made between the flat surfaces of the bonnet and the top of the body. A weld is then made around lips formed by

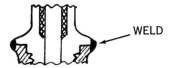

Fig. 2-14. Lip-sealed bonnet.

the bonnet-body joint. The weld acts only as a liquid seal while the threads are carrying the entire mechanical load. Since the weld is laid on a flat surface, it can be ground off in order to remove the bonnet for maintenance purposes.

Breech-Lock Bonnet

This is another style bonnet for use where high pressures and/or temperatures will be encountered. Figure 2-15 illustrates such an assembly. The thrust on the bonnet is transmitted to the body through heavy interlocking breech lugs. The valve is assembled by lowering the bonnet into the body and rotating the bonnet 45 degrees to engage the body and bonnet lugs. The assembly is completed by making a small seal weld between the body and a flexible steel ring on the bonnet. By chipping out the welded joints the valve may be disassembled without removing the body from the line.

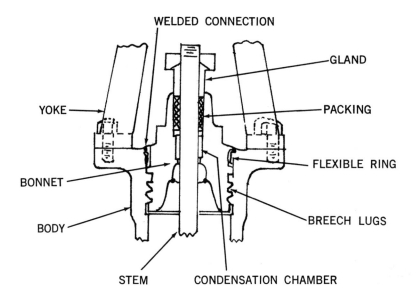

Fig. 2-15. Breech-lock bonnet.

2.4 METHODS OF VALVE OPERATION

Most valves are operated simply, by means of the handwheel or lever which is supplied with the valve. However, there are times when it is either inconvenient, undesirable, or impossible to operate the valve in this manner. In order to meet these needs there are a variety of alternative methods of manual operation and a variety of automatic operators. These items are considered as valve accessories and can be supplied to fit most types or styles of valves.

Accessories for Manual Operation

The accessories for manual operation are usually supplied to solve one of two problems, namely:

1. Inaccessibility of the valve, thus preventing manual operation in the normal manner.
2. The valve is of such a size that a conventional handwheel will not provide sufficient leverage to permit an average man to open and close the valve.

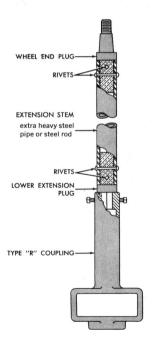

Courtesy of Jenkins Bros.

Fig. 2-16. Stem extension rod for rising stem valve.

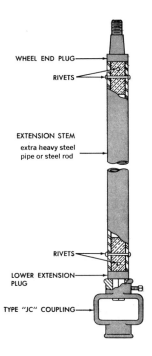

Fig. 2-17. Stem extension rod for nonrising stem valves.

Extension Stems

One of the most common problems encountered is the need to operate a valve which is located out of reach. It is not always possible or practical to place operating valves within reach. Consequently, valve manufacturers can and do supply extension stems.

Stem extension units permit remote operation of valves by providing a stem extension of any required length between the valve and the valve-operating mechanism such as a wheel, key, or floor stand. The extension units usually consist of a cold rolled steel rod and a coupling for attaching to the valve stem. For a longer extension the rod is replaced with a heavy steel pipe, securely fastened to a wheel end plug and a lower extension plug, as shown in Fig. 2-16. A square hole wheel or operating nut fits the tapered square tip of the rod or wheel end plug. Adequate and rigid support must be provided for extra long stem extension units. Alternatively, a flexible metal shaft may serve as an extension stem.

The stem extension unit shown in Fig. 2-16 is used on rising stem valves. It is secured to the yoke sleeve of the valve by the yoke sleeve nut. The upper portion is hollow to receive the rising stem of the valve.

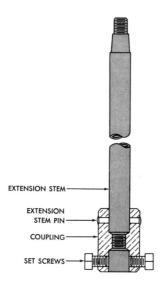

Fig. 2-18. Typical stem extension rod for valves through 3-inch size.

A typical stem extension unit, shown in Fig. 2-17, is used for non-rising stem valves. Figure 2-18 shows a typical stem extension unit for use on valves through 3-inch size. The lower hub of the extension stem is threaded to screw onto the valve stem and then secured to it by two set screws. The regular valve handwheel fits the squared end of the extension stem and is secured to it by the wheel nut.

In addition to the rigid stem extension units, other methods are also available to permit manual operation of valves which cannot be reached from operating levels. Such methods include the installation of flexible shafts from the valve stem to the operating level or the use of steel rods and universal joints. These latter two methods permit the operating location to be at any convenient point (within reason) not necessarily directly in line with the valve stem.

Floor Stands

Floor stands are designed for operating gate, globe, or other valves installed in pipe trenches or other inaccessible locations. Floor stands are usually supplied with a special coupling for connecting the stem extension unit to the operating stem of the floor stand. These floor stands are shown in Fig. 2-19, and as can be seen are available with or without indicator attachments. The indicator attachment shows the position of the flow control element.

VALVE DESIGN

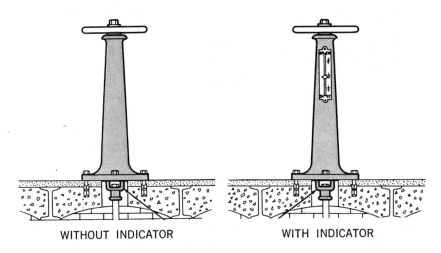

WITHOUT INDICATOR | WITH INDICATOR

Courtesy of Jenkins Bros.

Fig. 2-19. Typical floor stands for valves.

Courtesy of NIBCO Inc.

Fig. 2-20. Typical chain wheel operator.

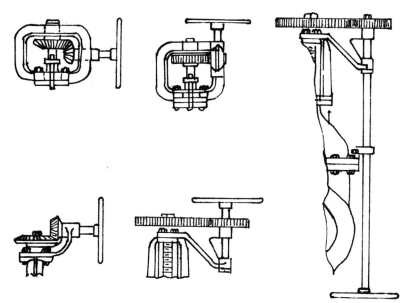

Fig. 2-21. Typical arrangements of gear operators.

Wheel Operators

Wheel operators are used to permit operation of overhead valves when an extension stem will not suffice, such as in a vertical pipeline or when the height makes an extension stem impractical. Figure 2-20 illustrates a wheel operator. These operators are easily attached to the rim of the valve wheel — *never to the spokes.* Adjust the hook bolts to the wheel rim, pass them through the slots in the lugs, center the sprocket rim on the wheel and tighten the nuts. When chain wheels are used on valves that have a handwheel mounted on the stem, make sure that the stem is strong enough to withstand the extra weight and pull. Valves can be ordered equipped with chain wheel operators.

Gear Operators

Gear operators are used to provide additional mechanical advantage in the opening and closing of larger valves. Operators can be mounted directly on the valve or remotely by means of extension stems. Several styles of gearing arrangements are shown in Fig. 2-21.

Hammer Blow Handwheel

Hammer blow handwheels are designed for use on large-size or high-pressure valves where additional force is necessary for the initial effort in starting the disc or wedge or for the final effort in effecting tight seating. The hammer action is provided by two driving lugs cast on the underside of the wheel and the lug of the anvil which projects into the wide slot between the driving lugs.

At the full open or closed positions one of the driving lugs strikes the anvil with sharp impact, facilitating positive opening or closing of the valve. The hammer blow handwheels are usually of large diameters (24 inches or more) and are supplied as standard on some valves.

Position Indicators

Position indicators are used on nonrising stem valves to show the relative position of the flow control device. A typical unit is found in Fig. 2-22. An indicator travels up or down in the slot as the valve is opened or closed showing the relative position of the flow control element.

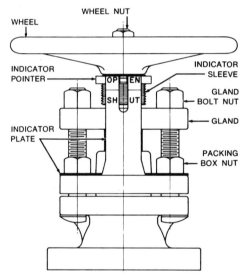

Courtesy of Jenkins Bros.

Fig. 2-22. Valve with position indicator.

Accessories for Automatic Operation

On occasion it may be necessary or desirable to have valves operate automatically. Such automatic operation may be of a throttling nature, or it can be a simple open and close operation. This can be accomplished by the addition to the standard valve of any of the following accessories:

1. Air (hydraulic) operator for throttling and/or open-closed operation

2. Electric solenoid for open-closed operation
3. Electric motor for throttling and/or open-closed operation.

Air (Hydraulic) Operators (Air Motors)

This type of operator, available with either a diaphragm or piston construction, is the most widely used. Regardless of style, the general method of operation is the same. In the diaphragm type two chambers are isolated one from the other by means of an elastomeric diaphragm. The diaphragm is connected to the valve stem, and as it is forced up or down it opens or closes the valve. The diaphragm can be positioned intermediately to permit a throttling action by the valve. The position is relative to the air pressure admitted to the valve operator.

In general, three types of air motors are available.

1. Air pressure to open — spring to close
2. Air pressure to close — spring to open
3. Air pressure to open and air pressure to close.

The reverse action of types 1 and 2 is accomplished by means of a spring which acts upon the diaphragm holding the valve in either the closed (type 1) or open (type 2) position. As air pressure is increased on the side of the diaphragm away from the spring, valve action takes place. In the event of an air failure the valve will return to its original open or closed position, which is usually referred to as the depressurized position. This is the valve position desired should there be a failure in the air supply. In this position no danger exists, and no damage can be caused as a result of the valve's position. This is also known as the "fail-safe" position.

Since the force generated to operate the valve is a function of the area of the diaphragm and the operating air pressure, the higher the operating air pressure the smaller the area of the diaphragm required and consequently the lower the cost of the air motor.

As indicated, operation can be of either the open-closed or throttling type. Open-closed type of operation can be effected by using electrically operated solenoid valves or manually operated 3- and 4-way valves to control the air flow (or hydraulic fluid) to the diaphragm chamber. When solenoid valves are ordered for hydraulic use, the fluid should be specified to insure that proper seals are included in the solenoid. To insure quick response of control, solenoid valves should be located as close as possible to the topworks they are controlling. Solenoid valves can be used in conjunction with either toggle switches, timers, pressure switches, pressure differential switches, temperature switches, and/or electrical relays.

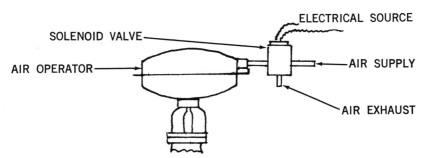

Fig. 2-23. Air operator (air-to-close, spring-to-open) arrangement using 3-way solenoid valve as pilot control.

A typical arrangement showing a 3-way solenoid valve used as a pilot control for an air-to-close, spring-to-open actuator can be found in Fig. 2-23.

Double-acting topworks (air-to-open air-to-close) can be controlled by two 3-way solenoids controlling the separate air chambers and wired to operate simultaneously, or they can be controlled by a single 4-way solenoid valve which admits air to one chamber while simultaneously exhausting the other to open or shut the valve. A typical arrangement is shown in Fig. 2-24.

In most applications the automatic throttling of flow is usually the result of an air signal supplied to the air motor from some form of controller. The purpose of control may be to maintain flow, pressure, temperature, liquid level, or other control requirement. A typical arrangement is illustrated in Fig. 2-25 showing an air motor with a spring-to-close and air-to-open operation. The controller regulates the air pressure

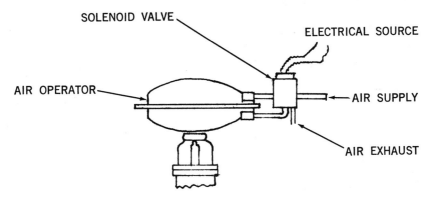

Fig. 2-24. Air operator (air-to-open, air-to-close) arrangement using 4-way solenoid valve.

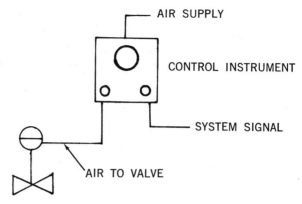

Fig. 2-25. Typical arrangement of automatic controller and air operator.

supplied to the valve motor to provide the proper valve opening so that control can be maintained over the system.

Electric Solenoid Operators

Solenoid valves are designed for open or closed operation. They are available as either normally open (NO) or normally closed (NC). This designation refers to the position of the valve when installed before the coil is energized. When the coil of a normally open valve is energized, the valve will close. In the event of a power failure the valve will open and remain open. The converse is true of a normally closed valve.

Since solenoid valves are quick acting, provision should be made in the piping system to accommodate the hydraulic shock accompanying the operation of these valves. If not compensated for, such hydraulic shock can seriously damage the piping system.

Electric Motor Operators

Electric motor operation is similar to air operation in that the valve opening can be set in intermediate positions. The valve stem is connected to a small electric motor which positions the valve.

2.5 VALVE CONNECTIONS

Valves are furnished with end connections to accommodate all of the various conventional methods of joining pipe. However, this does not mean that all valves, in all types of construction, in all sizes, are available with the complete range of end connections. For example, lined valves are usually only available in a flanged construction since it is not possible to provide a threaded end with most lining materials.

Threaded (Screwed) Ends

Threaded ends are usually provided with tapped female threads into which the pipe is threaded. (American Standard Pipe Threads ANSI B2.1). Threaded-end valves are the least expensive, as less material and less finishing are required. This is especially true in higher alloy materials of construction. In addition, they can be quickly and easily installed in a line since no special pipe end preparation (other than threading) is required. However, when installing a threaded-end valve, two pipe unions should be installed — one on either side of the valve, fairly close to the valve — to facilitate valve removal for maintenance purposes.

Flanged Ends

Flanged ends make a stronger, tighter, more leak-proof connection than threaded ends. In certain alloy construction, such as stainless steel, it can be difficult to make a tight, leak-free, threaded connection. Where heavy viscous media are to be controlled, flanged-end valves should be specified. Because of the additional metal required by the flanges and the accurate machining required on the flanges, these valves will have a higher initial cost. The installation cost will also be greater because companion flanges — for installation on the pipe ends to which the valve end flanges will be bolted — as well as gaskets, bolts, and nuts must also be provided.

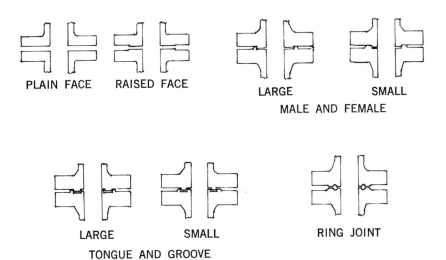

Fig. 2-26. Typical flange connections.

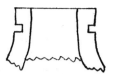

Fig. 2-27. Grooved-end connection.

The different types of flanged ends available are shown in Fig. 2-26. Bronze and iron flat-face flanges have a machined finish.

Steel flat-face flanges and raised face flanges have a serrated finish of approximately 32 serrations per inch and may be either spiral or concentric. Steel, male and female, and tongue and grooved faces have either an RMS 64 finish for liquids or an RMS-32 finish for gases. The correct finish should be specified when ordering. The steel ring joint faces have similarly finished grooves of either RMS-64 or RMS-32 specification.

Grooved Ends

The Victaulic Company of America provides a patented coupling for connecting grooved piping. Grooved end valves are available for installation in such a piping system. A typical end connection is shown in Fig. 2-27.

Since there are no end flanges, bolts, and nuts, the valves are light in weight and can be quickly and easily installed or taken out of the line. The only tool required is an ordinary socket or speed wrench.

In addition to the end configurations described, valves with other types of end connections are also available. Most of these are designed to be used in conjunction with specific types of pipe-coupling designs. Typical examples of such end configurations for use with the appropriate coupling would be Swagelok, Grayloc, Marmon, etc.

Solder End and Silver Brazing End

Valves are available with end connections to permit connection to pipe or tubing by means of soldering or silver brazing. Typical end connections are shown in Fig. 2-28. This provides a quick economical method of installation.

Socket or Butt Welding Ends

For installation in all welded piping systems valves are available with ends suitable for either socket or butt welding. Figure 2-29 shows typical valve ends for socket welding and butt welding.

SOLDER END   SILVER BRAZING END

Fig. 2-28. End connections for solder and silver brazing. Valves for brazing are also available containing preformed braze material inserts.

2.6 INSTALLATION TECHNIQUES

As stated previously, the most important decision in developing valve specifications is to fit the right valve to the right job. Second in sequence, but as important, is valve location, and third is valve installation. All three steps are equally important if long satisfactory service is to be obtained from the valves.

Valve Location

Valves should be so located in a pipeline that they can be easily and safely operated. If remote operation (either manual or automatic) is not to be employed, the valves should be located so that the operator can exert just the right amount of force to open and close them properly. Overhead valves, with the handwheels facing down, are usually installed so that the handwheel is at an elevation of 6 feet 6 inches above the

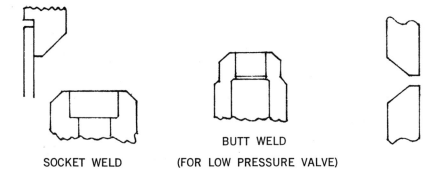

SOCKET WELD   BUTT WELD (FOR LOW PRESSURE VALVE)

Fig. 2-29. End connections for socket weld and butt weld joints.

operating floor. This places the handwheel high enough so as not to present a head hazard and low enough so that an operator of average height can reach the handwheel easily and safely.

If the valve is installed higher (e.g., at operator fingertip height), the operator is forced to stretch to reach the handwheel. In this position he cannot close the valve tightly, and eventually leakage occurs, which may cause abnormal wear on the seat and disc assemblies.

Valve Care Before Installation

Valves are generally wrapped and/or protected from damage during shipment by the manufacturer. This wrapping and/or protection should be left in place until the valve is to be installed. If the valves are left exposed, sand or other gritty material may get into the working parts. Unless all such foreign material is thoroughly cleaned out, it may cause trouble when the valve has been placed in service.

Valves should be stored where they are protected from corrosive fumes, and in such a manner that they will not fall or where other heavy material will not fall onto the valves.

Prior to installation it is advisable to have all valves either blown out with *clean* compressed air or flushed out with water to remove all dirt and grit. Piping should be cleaned out in the same manner, or swabbed out to remove dirt or metal chips left from threading operations or welding on the pipe.

Relieving Pipe Stresses

Pipe carrying high temperature fluids will be subject to thermal stresses due to the thermal expansion of the piping system. Unless provision is made to take up the expansion of the pipe length involved, these stresses will be transmitted to the valves and pipe fittings.

The expansion of the pipe can be accommodated by installing either a "U" expansion bend or an expansion joint between all anchor points. Whichever method is used, care must be taken to insure that there is sufficient movement available to accommodate the expansion of the length of pipe involved. Note that the same condition exists, but in the opposite direction, on pipelines carrying extremely cold fluids. In these instances pipe contraction must be compensated for.

Installation of Valves

This section deals with the general installation procedures to be followed for all types of valves. Unique procedures for specific types of valves will be included in the section dealing with those valves.

When installing valves, make sure that all pipe strains are kept off of the valves. The valves should not carry the weight of the line. Distortion from this cause results in inefficient operation, jamming, and the necessity of early maintenance. If the valve is of flanged construction, it will be difficult to tighten the flanges properly. Piping should be supported by hangers placed on either side of the valve to take up the weight. Large heavy valves should be supported independently of the piping system so as not to induce a stress into the piping system.

When installing rising stem valves, be sure to allow sufficient clearance for operation and for removal of the stem and bonnet if necessary. Insufficient clearance will prevent the valve from being fully opened, which will result in excessive pressure drop, gate wedge erosion, chatter and wire drawing or seat wear.

It is better to install valves with the stem in the upright position. However, most valves may be installed with the stem at any angle. When installed so that the stem is in the downward position, the bonnet is under the line of flow, forming a pocket to catch and hold any foreign matter. (An exception to this are valves with a diaphragm which seals the bonnet from the material in the pipeline.) Foreign material, so trapped, can eventually cut and ruin the inside stem or threads.

Threaded End Valves

Avoid undersize threads on the section of pipe where the valve is to be installed. If the threaded section of pipe is too small, the pipe, when screwed into the valve to make a tight connection, may strike the diaphragm and distort it so that the disc or wedge will not seat perfectly. Undersize threads on the pipe also make it impossible to get a tight joint. A safe practice is to cut threads to standard dimensions and standard tolerances. All pipe threads in valve bodies are tapered and gaged to standard tolerances.

Paint, grease, or joint sealing compound should be applied only to the pipe (male) threads — *not* to the threads in the valve body. This reduces the chances of the paint, grease, or compound getting on the seat or other inner working parts of the valve to cause future trouble.

When installing screwed end valves always use the correct size wrenches with flat jaws (not pipe wrenches). By so doing there is less likelihood of the valve's being distorted or damaged. Also, the wrench should be used on the pipe side of the valve to minimize the chance of distorting the valve body. This is particularly important when the valve is constructed of a malleable material, such as bronze. As a further precaution the valve should be tightly closed before installation.

Flanged End Valves

When installing flanged valves, tighten the flange bolts by pulling down the nuts diametrically opposite each other and in the order numbered as shown in Fig. 2-30.

All bolts should be pulled down gradually to a uniform tightness. Make all bolts finger tight first, then take three or four turns with a wrench on bolt 1. Apply the same number of turns on each bolt, following the order shown in Fig. 2-30. Repeat the procedure as many times as required until the joint is tight. Uniform stress across the entire cross section of the flange eliminates a leaky gasket.

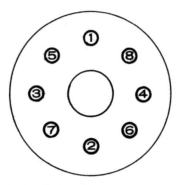

Fig. 2-30. Order of tightening flange bolts.

Socket Weld End Valves

The following procedure should be followed for the installation of valves having socket welding ends. It is recommended that the manufacturer be requested to weld 3-inch-long nipples in the valve bodies of all socket welding valves in sizes of 2 inches and smaller. This can be done before the interior of the valve is machined and reheat treated. By doing this, it will be possible to weld the nipples directly to the pipe and/or pipe fittings without distorting the valve body and seats.

1. Cut the pipe end square, making sure that the diameter is not undersize or out of round. Remove all burrs.

2. Clean pipe end (at least to depth of socket) and inside of socket with a degreasing agent to remove oil, grease, and foreign matter.

3. Insert pipe into valve socket and space as shown in Fig. 2-31 by backing off pipe 1/16 inch after it hits against the shoulder within the valve socket, or by using a removable spacing collar. This procedure is very important before welding. Tack weld in place.

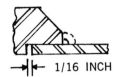

Fig. 2-31.   Positioning for socket weld.

4. Make certain that valve is in open position before applying heat. Valve bonnets should be hand tight to prevent distortion or damage to threads. Valves having nonmetallic discs should have same removed before heat is applied. Valve and pipe should be supported during socket welding process, and must not be strained while cooling.

5. Preheat welding area 400°F to 500°F.

6. For the highest-quality weld, the inert gas-arc method is recommended. Using the inert gas-arc method or metallic-arc method, a socket weld is normally completed in two or more passes. Make sure the first pass is clean and free from cracks before proceeding with the second pass. *Excessive heat causes distortion and improper functioning of the valve.*

7. Discoloration may be removed by wire brushing.

Butt Weld End Valves

It is recommended that the manufacturer be requested to weld 3-inch long nipples on the valve bodies of all butt welding end valves in sizes 2 inches and smaller. This can be done before the interior of the valve is machined and reheat treated. By doing this it will be possible to weld the nipples directly to the pipe and/or pipe fittings without distorting the valve body and seats. The following procedure should be used for the installation of valves having butt welding ends.

1. Machine pipe ends for butt welding joint. Remove all burrs.

2. Clean pipe and valve joints with a degreasing compound to remove oil, grease, and all foreign matter.

3. Space joint as shown in Fig. 2-32. Align by means of fixtures. Tack weld in place.

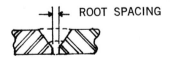

Fig. 2-32.   Positioning for butt welding.

4. Make certain valve is in open position before applying heat. Valve bonnet should be hand tight to prevent distortion or damage to threads, and valves having nonmetallic discs should have same removed before heat is applied. Valve and pipe should be supported during butt-welding process, and must not be strained while cooling.

5. Preheat welding area 400°F to 500°F.

6. For the highest quality weld, the inert gas-arc method is recommended. Using either the inert gas-arc method or metallic-arc method, a butt weld is normally completed in two or more passes. The first pass should have complete joint penetration and be flush with the internal bore of the pipe. Make sure that the first pass is clean and free from cracks before proceeding with the second pass. The final pass should blend with the base metal and be flush with the external diameter. Indication of cracks, lack of fusion, or incomplete penetration are causes for rejection. *Excessive heat causes distortion and improper functioning of the valve.*

7. Discoloration may be removed by wire brushing.

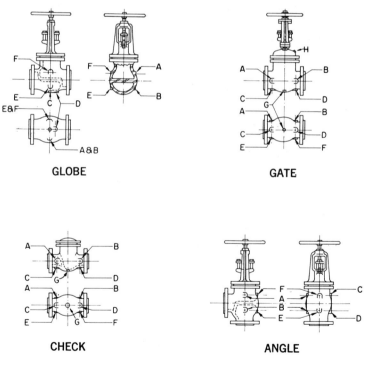

Courtesy of NIBCO Inc.

Fig. 2-33. Boss locations.

Ch. 2   VALVE DESIGN   45

8. After completion of the weld it should be subjected to one of the methods of nondestructive testing to insure a sound weld, with complete penetration data. Detailed information regarding the various methods of nondestructive testing may be gotten from other sources dealing with these topics.

2.7 DRAIN CONNECTIONS AND BYPASSES

A convenient means of draining a section of pipe between two block valves is to supply the valves with drain connections. Most bronze valves are available with this optional feature which consists of the installation, by the manufacturer, of a ⅛-inch drain cap.

Iron body valves are available with bosses which can be tapped in accordance with MSS Bypass and Drain Connection Standard SP-45. Standard tapping sizes are shown in Table 2-1.

Table 2-1. Standard Size of Tapping

Size of Valve Inches	2	3	4	5	6	8	10	12
Size of Drain Tapping Inches	½	½	½	¾	¾	¾	1	1

Boss locations are standardized as are the boss symbols. These are shown in Fig. 2-33.

Bypasses are utilized to equalize pressure at inlet and outlet before opening main valve, thus facilitating easy valve operation. They may also be used for preheating the outlet lines and eliminate damage from too fast expansion. The standard bypass sizes, in accordance with MSS specification SP-45 Table 11 Series A for steam service are shown in Table 2-2.

Table 2-2. Sizes of Bypass Connections

Main Valve Size, Inches	4	5	6	8	10	12
Bypass Valve Size, Inches	½	¾	¾	¾	1	1

3

Flow Through Valves

In the design of a piping system, of which valves are a critical part, an important engineering problem is the calculation of the loss of energy due to the resistance of fluids to movement through pipe or closed channels. The Bernoulli Theorem provides a general energy equation which takes into account all of the factors, by application of the law of conservation of energy to the flow of fluids in a conduit.

The total energy at any point taken above an arbitrary horizontal datum plane in any piping system is equal to the sum of the elevation head (static head), the pressure head, and the velocity head. It is expressed as follows:

$$H = Z + \frac{144P}{\rho} + \frac{v^2}{2g} \tag{3-1}$$

where:

H = Total head of the system, in feet
Z = Potential head or elevation above reference level, in feet
$\frac{144P}{\rho}$ = Pressure head in feet
P = Pressure of system in pounds/square inch
ρ = Fluid density in pounds/cubic foot
$\frac{v^2}{2g}$ = Velocity head in feet
v = Fluid velocity in feet/second
g = Acceleration of gravity = 32.2 feet/second/second

All practical formulas used to calculate fluid flow and/or head losses (pressure drops) are derived from the Bernoulli Theorem with modifications to account for losses resulting from friction.

FLOW THROUGH VALVES

Flow in a pipe is always accompanied by friction of fluid particles rubbing against one another and the wall of the pipe, and consequently by loss of energy or a pressure drop. If ordinary Bourdon tube pressure gauges were connected to a pipe containing a flowing fluid, the upstream pressure gauge would indicate a higher pressure than the downstream pressure gauge.

Many empirical formulas have been developed to determine the frictional losses for specific fluids under specific flow conditions. Most of these formulas follow the general equation:

$$h = f \frac{Lv^x}{d^y} \tag{3-2}$$

where:
h = Loss of head
f = Friction factor, based on experimental data
L = Length of pipe
d = Inside diameter of pipe
v = Average fluid velocity
x = Coefficient based on experimental data
y = Coefficient based on experimental data

Experimental work has shown that the friction factor is a function of the pipe diameter, average flow velocity, density and absolute viscosity of the fluid. This is shown by the equation:

$$f = \Phi \frac{Dv\rho}{\mu_a} \tag{3-3}$$

where:
D = Inside diameter of pipe, in feet
v = Average velocity, in feet per second
ρ = Density, in pounds/cubic foot
μ_a = Absolute viscosity in pounds/foot second
Φ = Function of flow conditions

The factor $\left(\dfrac{Dv\rho}{\mu}\right)$ is generally known as the Reynolds number (R_e).

For engineering purposes flow in pipes is usually considered to be viscous (laminar) if the Reynolds number is less than 2000 and turbulent if the Reynolds number is greater than 4000. Between these two values lies a critical zone where the flow is unpredictable because it may be laminar, turbulent, or in the process of change depending upon many variables.

The friction factor for laminar flow conditions ($R_e < 2000$) is a function of the Reynolds number only, whereas for turbulent flow conditions ($R_e > 4000$) it is also a function of the nature of the pipe wall.

The general equation for pressure drop in a pipe, known as Darcy's formula and expressed in feet of liquid is:

$$h_L = \frac{fLv^2}{D2g} \qquad (3\text{-}4)$$

where:

h_L = Loss of static pressure due to fluid flow, in feet of liquid
f = Friction factor
L = Length of pipe, in feet
D = Internal diameter of pipe, in feet
v = Mean velocity of flow, in feet per second
g = Acceleration of gravity = 32.2 feet per second per second

This equation may also be written to express the pressure drop in pounds per square inch as follows:

$$\Delta P = \frac{fLv^2\rho}{144D2g} \qquad (3\text{-}5)$$

where:

ΔP = Loss of pressure in pounds per square inch

If the flow is laminar ($R_e < 2000$), the friction factor may be calculated from the equation:

$$f = \frac{64\mu}{124dv\rho} \qquad (3\text{-}6)$$

If this quantity is substituted into equation 3-5, the pressure drop in pounds per square inch is:

$$\Delta P = \frac{(6.68 \times 10^{-4})\mu Lv}{d^2} \qquad (3\text{-}7)$$

where:

d = Inside diameter of pipe in inches

When the flow is turbulent ($R_e > 4000$), the friction factor depends not only on the roughness of the pipe but also on the size of the pipe. Since the character of the internal surface of commercial pipe is practically independent of the diameter, the roughness of the walls has a greater effect on the friction factor in the smaller diameters. In general, pipes of small diameters will have a higher friction factor than pipes of larger diameters in the same material of construction. The material of construction of the pipes will also be a factor since different materials have varying degrees of "roughness." For example, brass pipe is "smoother" than iron pipe. Friction factors for various pipe materials are usually available from the pipe manufacturers.

Pressure drop through valves is affected by the same variables as the pressure drop through a straight pipe. Because of this it is possible to express the resistance of valves as being equivalent to a length of straight pipe, and the approximate pressure drop for the actual flow conditions can be calculated by using the formulas developed for determining pressure drop through straight pipe. Experimental data has shown that the pressure drop through a valve can be expressed by the equation:

$$h_L = \frac{Kv^2}{2g} \qquad (3\text{-}8)$$

where:
  K = Resistance coefficient

By combining equation 3-8 with Darcy's formula (equation 3-4),

$$\frac{Kv^2}{2g} = h_L = \frac{fLv^2}{D2g}$$

the pressure drop of a valve can be converted to an equivalent length of straight pipe. Since:

$$K = \frac{fL}{D} \qquad (3\text{-}9)$$

it follows that

$$L = \frac{KD}{f} \qquad (3\text{-}10)$$

Theoretically the resistance coefficient K would be a constant for all sizes of a given design of valve if all sizes were geometrically similar. Since valve design is dictated by manufacturing economies, structural strength, standards, and other considerations, geometric similarity is seldom if ever achieved.

Numerous tests have been performed by valve manufacturers, universities, and others to determine the K coefficients. An analysis of these test results indicates that the slope of the K curves show a definite tendency to follow the same slope as the fL/D curve for straight pipe. Based on this it can be said that the resistance coefficient K, for a given type of valve, tends to vary with size as does the friction factor f for straight pipe and that the equivalent length L/D tends toward a constant for the various sizes of a given line of valves.

Usually the data of valve resistance to flow is expressed in equivalent feet of schedule-40 steel pipe.

To assist in making average calculations Table 3-1 provides the approximate flow resistance of fully opened valves compared to equivalent feet of clean schedule-40 steel pipe. In general, gate valves offer the least resistance to fluid flow, globe valves the greatest resistance, with the other types of valves falling somewhere in-between.

Table 3-1. Resistance to Flow of Fully Opened Valves Expressed in Equivalent Lengths (in Feet) of Clean Schedule-40 Steel Pipe

Valve Size Inches	Valve Type				
	Gate	Y	Swing Check	Angle	Globe
½	0.4	3.6	4	8	16
¾	0.5	4.5	5	12	22
1	0.6	6.3	7	15	27
1¼	0.8	8.0	9	18	37
1½	0.9	10.0	11	21	44
2	1.2	12.5	14	28	55
2½	1.4	13.5	15	32	65
3	1.6	18.0	19	41	80
3½	2.0	20.0	22	50	100
4	2.2	22.5	25	55	120
5	2.9	28.0	32	70	140
6	3.5	36.0	40	80	160
8	4.5	45.0	50	110	220
10	5.5	58.0	65	140	280
12	6.5	68.0	75	160	340
14	8.0	81.0	90	190	380
16	9.0	95.0	105	220	430
18	10.0	108.0	120	250	500
20	12.0	117.0	130	270	550
24	14.0	135.0	150	380	650

It has been found convenient in some branches of the valve industry, particularly in the area of control valves, to express the valve capacity in terms of the flow coefficient C_v. The C_v coefficient of a valve is defined as the flow of water at 60°F, in gallons per minute at a pressure drop of one pound per square inch across the valve. By substitution of appropriate equivalent units in the Darcy equation it can be shown that:

$$C_v = \frac{29.9d^2}{\sqrt{f\frac{L}{D}}} = \frac{29.9d^2}{\sqrt{K}} \quad (3\text{-}11)$$

Ch. 3　FLOW THROUGH VALVES

In addition the quantity Q in gallons per minute of any liquid having a viscosity close to that of water at 60°F that will flow through the valve can be determined from:

$$Q = C_v \sqrt{\Delta P \frac{62.4}{\rho}}$$

$$Q = 7.9 C_v \sqrt{\frac{\Delta P}{\rho}}$$
(3-12)

where:

ρ = Density of fluid in pounds per cubic foot
ΔP = Pressure drop across the valve ($P_1 - P_2$), in pounds per square inch

In a like manner the pressure drop can be calculated by rearranging the same formula as follows:

$$P = \frac{\rho}{62.4}\left(\frac{Q}{C_v}\right)^2$$
(3-13)

The C_v values are usually supplied by the valve manufacturers.

Since equations 3-8, 3-9, 3-11, 3-12, and 3-13 are simply modified forms of the Darcy equation, the limitations regarding their use for compressible flow apply as the following restrictions apply for the Darcy formula:

1. If the calculated pressure drop ($P_1 - P_2$) is less than about 10 percent of the inlet pressure P_1, reasonable accuracy will be obtained if the specific volume used in the formula is based upon either the upstream or downstream conditions, whichever are known.

2. If the calculated pressure drop ($P_1 - P_2$) is greater than 10 percent, but less than 40 percent of the inlet pressure P_1, the Darcy equation may be used with reasonable accuracy by using a specific volume based upon the average of upstream and downstream conditions; otherwise the method given in restriction 3 may be followed.

3. The maximum velocity of a compressible fluid in a pipe is limited by the velocity of propagation of a pressure wave which travels at the speed of sound in the fluid. This maximum possible velocity is sonic velocity which is expressed as:

$$v_s = \sqrt{kgRT} = \sqrt{kg144p'\overline{V}}$$
(3-14)

where:

v_s = Sonic (or critical) velocity of flow of a gas, in feet per second
k = Ratio of specific heat at constant pressure to specific heat at constant volume = C_p/C_v

R = Individual gas constant = $MR/M = 1544/M$ (M = molecular weight)
T = Absolute temperature in degrees Rankine ($460 + t°F$)
p' = Pressure, pounds per square inch absolute (gage pressure + barometric pressure)
$\overline{V}$ = Specific volume of fluid in cubic feet per pound

For most diatomic gases the value of $k = 1.4$, which is the ratio of specific heats at constant pressure to constant volume. This velocity will occur at the outlet end or in a constricted area, when the pressure drop is sufficiently high. The pressure, temperature, and specific volume are those occurring at the point in question.

SUMMARY OF FORMULAS

This summary of formulas[1] presents the basic formulas found throughout the chapter in terms of units which are most commonly used. This will provide the reader with an equation which will enable him to arrive at a solution to his problem with a minimum conversion of units.

To prevent duplication, formulas have been written in terms of either specific volume $\overline{V}$ or weight density, but not in terms of both, since one is the reciprocal of the other.

$$\overline{V} = \frac{1}{\rho} \qquad \rho = \frac{1}{\overline{V}}$$

Nomenclature used in these formulas will be found on page 56.

Reynolds Number of Flow in Pipe and Valves

$$R_e = \frac{Dv\rho}{\mu_e} = \frac{Dv\rho}{32.2\mu_e} = 123.9\frac{dv\rho}{\mu}$$

$$R_e = 22{,}700\frac{q\rho}{d\mu} = 50.6\frac{Q\rho}{d\mu} = 35.4\frac{B\rho}{d\mu}$$

$$R_e = 6.31\frac{W}{d\mu} = 0.482\frac{q'_h S_g}{d\mu}$$

$$R_e = \frac{Dv}{\nu'} = \frac{dv}{12\nu'} = 7740\frac{dv}{\nu}$$

$$R_e = 1.419 \times 10^6 \frac{q}{vd} = 3160\frac{Q}{vd} = 394\frac{W\overline{V}}{vd}$$

[1] By permission from Engineering Division, Technical Paper 410, Crane Co., Chicago, Illinois, 1970.

Viscosity Equivalents

$$\nu = \frac{\mu}{\rho'} = \frac{\mu}{S}$$

Darcy's Formula (Equation 3-4)

$$h_L = f\frac{L}{D}\frac{v^2}{2g} = 0.1863\frac{fLv^2}{d}$$

$$h_L = 6260\frac{fLq^2}{d^5} = 3.11 \times 10^{-2}\frac{fLQ^2}{d^5}$$

$$h_L = 1.524 \times 10^{-2}\frac{fLB^2}{d^5} = 4.83 \times 10^{-4}\frac{fLW^2\overline{V}^2}{d^5}$$

$$\Delta P = 1294 \times 10^{-3}\frac{fL\rho v^2}{d} = 3.59 \times 10^{-7}\frac{fL\rho v^2}{d}$$

$$\Delta P = 43.5\frac{fL\rho q^2}{d^5} = 2.16 \times 10^{-4}\frac{fL\rho Q^2}{d^5}$$

$$\Delta P = 10.58 \times 10^{-5}\frac{fL\rho B^2}{d^5} = 3.36 \times 10^{-6}\frac{fLW^2\overline{V}}{d^5}$$

$$\Delta P = 7.26 \times 10^{-9}\frac{fLT(q'h)^2 Sg}{d^5 P'}$$

$$\Delta P = 19.59 \times 10^{-9}\frac{fL(q'h)^2 Sg}{d^5}$$

Head Loss and Pressure Drop with Laminar Flow

If the flow is laminar ($R_e < 2000$), the friction factor may be calculated from equation 3-6:

$$f = \frac{64\mu}{124dv\rho}$$

The Darcy formula can be written as below when this value of f is substituted:

$$h_L = 9.62 \times 10^{-2}\frac{\mu Lv}{d^2}$$

$$h_L = 17.65\frac{\mu Lq}{d^4 \rho} = 3.93 \times 10^{-2}\frac{\mu LQ}{d^4 \rho}$$

$$h_L = 2.75 \times 10^{-2}\frac{\mu LB}{d^4 \rho} = 4.9 \times 10^{-3}\frac{\mu LW}{d^4 \rho^2}$$

$$\Delta P = 6.68 \times 10^{-4} \frac{\mu L v}{d^2} = 12.25 \times 10^{-2} \frac{\mu L q}{d^4}$$

$$\Delta P = 2.73 \times 10^{-4} \frac{\mu L Q}{d^4} = 1.91 \times 10^{-4} \frac{\mu L B}{d^4}$$

$$\Delta P = 3.4 \times 10^{-5} \frac{\mu L W}{d^4 \rho}$$

Nomenclature

B = Rate of flow in barrels (42 gallons) per hour
C_v = Flow coefficient for valves — expresses flow rate in gallons per minute of 60°F water with a 1.0 psi pressure drop across valve = $Q \sqrt{\dfrac{\rho}{62.4 P}}$
D = Internal diameter of pipe, in feet
d = Internal diameter of pipe, in inches
f = Friction factor
g = Acceleration of gravity 32.2 feet per second/per second
H = Total head, in feet of fluid
h = Static pressure head existing at a point, in feet of liquid
h_L = Loss of static pressure head due to fluid flow, in feet of liquid
K = Resistance coefficient
k = Ratio of specific heat at constant pressure to specific heat at constant volume = $\dfrac{C_p}{C_v}$
L = Length of pipe, in feet
M = Molecular weight
MR = Universal gas constant = 1544
P = Pressure, in pounds per square inch gage
p' = Pressure in pounds per square foot absolute
Q = Rate of flow, in gallons per minute
q = Rate of flow, in cubic feet per second
q' = Rate of flow, in cubic feet per second at standard conditions (14.7 psia and 60°F)
q'_h = Rate of flow, in cubic feet per hour at standard conditions (14.7 psia and 60°F) scfh.
R = Individual gas constant = $\dfrac{MR}{M} = \dfrac{1544}{M}$
R_e = Reynolds number
S = Specific gravity of liquids relative to water both at standard temperature (60°F)

Ch. 3 FLOW THROUGH VALVES 55

S_g = Specific gravity of a gas relative to air = the ratio of the molecular weight of the gas to that of air
T = Absolute temperature in degrees Rankine (460 + t)
t = Temperature in degrees Fahrenheit
$\overline{V}$ = Specific volume of fluid, in cubic feet per pound
V = Mean velocity of flow, in feet per minute
V_a = Volume, in cubic feet
v = Mean velocity of flow, in feet per second
v_s = Sonic (or critical) velocity of flow of a gas, in feet per second
W = Rate of flow, in pounds per hour
w = Rate of flow, in pounds per second
Z = Potential head or elevation above reference level, in feet
Δ = Differential between two points
ρ = Weight density of fluid, pounds per cubic foot
ρ' = Density of fluid, grams per cubic centimeter
μ = Absolute viscosity in centipoise
μ_e = Absolute viscosity, in pound mass per foot second poundal seconds per square foot
μ'_e = Absolute viscosity, in slugs per foot second or pound force seconds per square foot
ν = Kinematic viscosity, in centistokes
ν' = Kinematic viscosity, square feet per second

Typical Problems[2]

I. Given: A 6-inch 125-pound Y-pattern globe valve has a flow coefficient C_v of 600.

Find: The resistance coefficient K and the equivalent length (L) of 6-inch schedule-40 pipe.

1. $$C_v = \frac{29.9 d^2}{\sqrt{K}} \qquad (3\text{-}11)$$

$$K = \frac{891 d^4}{(C_v)^2}$$

2. For 6-inch schedule-40 steel pipe

d = 6.065 inches

D = 0.5054 feet

d^4 = 1352.8

[2] By permission from Engineering Division, Technical Paper 410; Crane Co., Chicago, Illinois, 1970.

3. $K = \dfrac{891(1352.8)}{(600)^2} = 3.35$

4. $K = \dfrac{fL}{D}$ (3-9)

from pipe friction factor chart $f = 0.015$

$$L = \frac{KD}{f} = \frac{(3.35)(0.5054)}{0.015}$$

$L = 112.8$ feet

II. Determine the pressure drop through a valve when 40 gallons per minute of a fluid having a density of 68 pounds per cubic foot are flowing through. The C_v of the valve is 50.

$$\Delta P = \frac{\rho}{62.4}\left(\frac{Q}{C_v}\right)^2 \qquad (3\text{-}13)$$

$$\Delta P = \frac{68}{62.4}\left(\frac{40}{50}\right)^2$$

$\Delta P = 0.7$ psig

4

General Purpose Valves

This section deals with the more common valves which are used to control flow in a piping system. Control of flow, as used here, refers to fully open, fully closed, or throttling services. Included in this category are the following valves:

Gate	Y-valves	Diaphragm
Globe	Plug	Pinch
Angle	Ball	Slide
Needle	Butterfly	

Table 4-1. Recommended Valve Services

Valve	Services							
	On-Off	Throt-tling	Divert-ing Flow	Freq. Oper.	Low Press. Drop	Slurry Han-dling	Quick Opening	Free Drain-ing**
Gate	×				×		×	×
Globe	×	×*		×				
Plug	×	×*	×	×	×		×	×
Ball	×	×*	×	×	×		×	
Butterfly	×	×		×	×	×	×	×
Diaph.	×	×*				×	×*	×*
Y	×	×*		×				
Needle		×						
Pinch	×	×			×	×		×
Slide	×				×	×	×*	×

* Certain configurations only. Check detailed section.
** All of these valves may not be completely free draining, but they trap a minimum amount of fluid in the line.

Section 5 is devoted to the so-called "special service" valves, which include:

Check valves Flush bottom tank valves
Relief valves Sample valves

Table 4-1 supplies the recommended services for the various valves discussed in Section 4. This will serve as a preliminary guide as to the type of valve to be considered for a specific service. It must be remembered that this is an overall tabulation and that the pages dealing with each specific valve should be consulted.

Table 4-2 summarizes the size ranges and operating ranges of the conventional valves. The ranges shown are average and will vary between manufacturers. Maximum figures are shown, which may not be attainable in all materials of construction. The same applies to the size ranges.

Table 4-2. Size and Operating Ranges of Valves

Valve	Sizes-Inches		Operating Ranges			
			Temperature, °F		Pressure, psi	
	Min.	Max.	Min.	Max.	Min.	Max.
Gate	1/8	48	−455	1250	Vacuum	10,000
Globe	1/8	30	−455	1000	Vacuum	10,000
Plug, lubricated	1/4	30	−40	600	Atm.	5,000
Plug, nonlubricated	1/4	16	−100	425	Atm.	3,000
Ball	1/4	36	−65	575	Atm.	7,500
Butterfly	2	36	−20	1000	Vacuum	1,200
Diaphragm	1/8	24	−60	450	Vacuum	300
Y	1/8	30	−455	1000	Vacuum	2,500
Needle	1/8	1	−100	500	Vacuum	10,000
Pinch	1	12	−100	550	Vacuum	300
Slide	2	75	0	1200	Atm.	400

4.1 GATE VALVES

General

Gate valves, being the simplest in design and operation of any valve, are the most widely used. The significant feature of this type of valve is less obstruction to flow, with less turbulence within the valve and very little pressure drop. When the valve is wide open, the wedge or gate is lifted entirely out of the waterway, providing a straight-way flow path through the valve. Gate valves offer less resistance to flow because the seating is at right angles to the line of fluid flow.

These valves are normally used where operation is infrequent and the valve will be either fully closed or fully opened. They should not be used for throttling operations. Except in the fully opened or fully closed position the gate and seat have a tendency to erode rapidly. Close control of flow is practically impossible because a great percentage of the flow change occurs near shut-off at high velocity. If the valve is opened slightly in a throttle position, the seat and disc are subjected to severe wire drawing and erosion that will eventually prevent tight shut-off.

Service Recommendations

1. Fully opened or fully closed, nonthrottling service
2. Minimum resistance to flow
3. Minimum amount of fluid trapped in line
4. Infrequent operation.

Construction of Valve

Gate valves are available with a variety of fluid control elements. Classification of gate valves is usually made by the type of fluid control element used. These elements are available as:

1. Solid wedge
2. Flexible wedge
3. Split wedge
4. Double disc (parallel disc).

Solid wedges, flexible wedges, and split wedges are employed in valves having inclined seats, while the double discs are used in valves having parallel seats.

Regardless of the style of wedge or disc used, they are all replaceable. In services where solids or high velocity may cause rapid erosion of the seat or disc, these components should have a high surface hardness and should have replaceable seats as well as discs. If the seats are not replaceable, damage to seats would require removal of the valve from the line for refacing of the seat, or refacing of the seat in place. Valves being used in corrosive service should always be specified with renewable seats.

Solid Wedge

The solid, or single wedge type, shown in Fig. 4-1 is recommended for use with oil, gas, air, and is preferred for slurries and heavy liquids. It can also be used for steam service where a double or split disc would

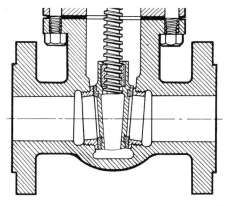

Courtesy of NIBCO Inc.

Fig. 4-1. Solid wedge.

result in excessive rattle or chatter, although the flexible disc is better for this service. These valves can be installed in any position.

Flexible Wedge

Flexible wedges (discs) are of one piece construction as shown in Fig. 4-2. The disc instead of being made completely solid with both seating surfaces rigid, is flexible. Thermal expansion and contraction entail no problems, as the flexible wedge is able to compensate for this and remain easy to open. These valves can be installed in any position.

Split Wedge

Split wedges, as shown in Fig. 4-3, are of the ball and socket design which are adjusting and self aligning to both seating surfaces. The disc

Fig. 4-2. Flexible wedge.

Ch. 4   GENERAL PURPOSE VALVES   61

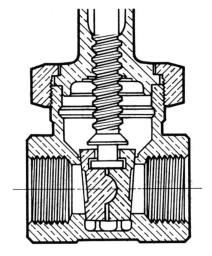

Courtesy of NIBCO Inc.

Fig. 4-3. Split wedge.

is free to adjust itself to the seating surface if one-half of the disc is slightly out of alignment because of foreign matter lodged between the disc half and the seat ring. This type of wedge (disc) is suitable for handling noncondensing gases and liquids at normal temperatures, particularly corrosive liquids. Freedom of movement of the discs in the carrier prevents binding even though the valve may have been closed when hot and later contracted due to cooling. This type of valve should be installed with the stem in the vertical position.

Double Disc

The double disc, or parallel disc, consists of two discs that are forced apart against parallel seats at the point of closure by a spreader device. See Fig. 4-4. This provides tight sealing without the assistance of fluid

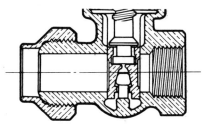

Courtesy of NIBCO Inc.

Fig. 4-4. Double disc.

pressure. The gate assembly automatically compensates for angular misalignment of the seats or longitudinal shrinkage of the valve body on cooling. Applications are much the same as for split wedges; i.e., for noncondensing gases, corrosive liquids, and light oil services. These valves are recommended for installation in horizontal lines with the valve stem in the vertical position.

Seats

Seats for gate valves are either provided integral with the valve body, or in a seat ring type of construction. Seat ring construction provides seats which are either threaded into position or pressed into position and seal-welded to the valve body. The latter form of construction is recommended for higher temperature service.

Integral seats provide a seat of the same material of construction as the valve body while the pressed-in or screwed-in seats permit variation. Rings with hard facings may be supplied for the applications where they are required.

Screwed-in rings are considered replaceable since they may be removed and new seal rings installed.

Stem Assemblies

Gate valves are available with stem assemblies of the following types:

1. Inside screw rising stem
2. Outside screw rising stem
3. Inside screw nonrising stem
4. Sliding stems.

For a discussion of the advantages, disadvantages, and recommended applications of each type of stem assembly, refer to pages 14 through 17.

Stem Seals

Various methods are available to prevent leakage along the valve stem to the outside of the valve. These are discussed starting on page 18; however, gate valves are usually supplied with packing boxes.

Bonnet Construction

Gate valves are available with a variety of bonnet constructions. Those generally available are:

1. Screwed-in bonnet
2. Screwed-on bonnet
3. Union bonnet
4. Flanged bonnet

5. U-bolt bonnet
6. Pressure sealed bonnet
7. Lip sealed bonnet
8. Breech lock bonnet

A discussion of each of these types of bonnets will be found starting on page 22.

Identification of Parts

The nomenclature and identification of the different components and parts of gate valves with varying bonnet construction, stem assemblies, and styles of wedges are shown in **Fig. 4-5**.

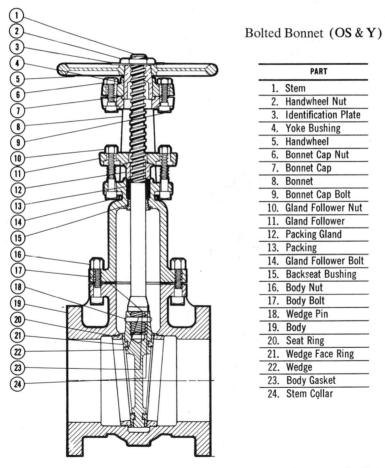

Bolted Bonnet (OS & Y)

PART
1. Stem
2. Handwheel Nut
3. Identification Plate
4. Yoke Bushing
5. Handwheel
6. Bonnet Cap Nut
7. Bonnet Cap
8. Bonnet
9. Bonnet Cap Bolt
10. Gland Follower Nut
11. Gland Follower
12. Packing Gland
13. Packing
14. Gland Follower Bolt
15. Backseat Bushing
16. Body Nut
17. Body Bolt
18. Wedge Pin
19. Body
20. Seat Ring
21. Wedge Face Ring
22. Wedge
23. Body Gasket
24. Stem Collar

Courtesy of NIBCO Inc.

Fig. 4-5. Identification of gate valve parts.

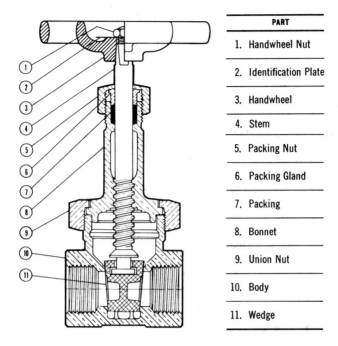

PART
1. Handwheel Nut
2. Identification Plate
3. Handwheel
4. Stem
5. Packing Nut
6. Packing Gland
7. Packing
8. Bonnet
9. Union Nut
10. Body
11. Wedge

Non-Rising Stem, Screw-in Bonnet

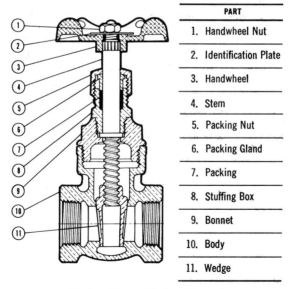

PART
1. Handwheel Nut
2. Identification Plate
3. Handwheel
4. Stem
5. Packing Nut
6. Packing Gland
7. Packing
8. Stuffing Box
9. Bonnet
10. Body
11. Wedge

Rising Stem, Union Bonnet

Fig. 4-5 (*continued*). Identification of gate valve parts.

Available Materials of Construction

Gate valves are available in a wide range of materials of construction. Standard representative constructions include:

1. Bronze
2. All iron
3. Cast iron
4. Forged steel
5. Monel
6. Cast steel
7. Stainless steel
8. Polyvinyl chloride (PVC)

Various trims are available on the above valves. Other metallic constructions are also available, some as standards and some on special order, depending upon the specific manufacturer.

Installation, Operation, and Maintenance

General installation procedures and care of valves prior to installation are given starting on page 40. Covered here are the specified items unique to gate valves.

Gate valves should only be used in services where they can always be either fully opened or fully closed. The bottom of the wedge (disc) or gate and seal will erode very rapidly if the gate is left in an intermediate position. In addition the wedge will tend to chatter and cause noise in the line.

Valves should be opened slowly to prevent hydraulic shock in the line. Closing the valve slowly will help to flush trapped sediment and dirt. Never force a valve closed with wrench or pry. When the valve has been fully opened, rotate the handwheel one quarter turn toward the closed position so as not to leave the handwheel jammed in the open position.

When closing down a "hot" line, wait until the system has cooled, then recheck closed valves to insure shut-off.

Replacing wedges in a gate valve will rarely cure a leak because the seat will have worn comparably to the wedge. When reseating a gate valve, be sure to mark the disc so that the disc is inserted in the valve body the same way it was removed. Otherwise a tight closure may not be obtained.

Packing leaks should be corrected immediately by tightening the packing nut which compresses the packing. If left unattended long enough, corrosive fluids will ruin the stem. In addition, a leaking stem can lead to valve chatter in which the vibration will damage the working parts of the valve. This is true on both small and large-sized valves. If it is apparent that the packing gland has compressed the packing to its limit, replace with new packing. Make sure that the packing used is resistant to the materials being handled by the valve.

Lubrication of valves is especially important and should be done on an established schedule. Valves that are opened and closed frequently should be lubricated at least once a month. On O.S. & Y. valves, when the stem is exposed, the screw threads should be kept clean and lubricated. Many valves are equipped with a lubricant fitting in the upper yoke to utilize pressure lubricant gun operation. Stem threads left dry and unprotected will become worn by grit and other abrasives threatening stem failure.

Gate Valve Specifications

Once it has been decided that gate valves are to be used, a specification should be established that will provide a valve which will meet all of the requirements of the system. This specification should include the following items as well as the materials of construction of the various components:

1. Type of end connection
2. Type of wedge
3. Type of seat
4. Type of stem assembly
5. Type of bonnet assembly
6. Type of stem packing
7. Pressure rating (operating and design)
8. Temperature rating (operating and design).

4.2 GLOBE VALVES, ANGLE VALVES, NEEDLE VALVES

General

Globe valves are normally used where operation is frequent and/or primarily in throttling service to control the flow to any desired degree. The significant feature of this valve is efficient throttling, with minimum wire drawing or disc and seat erosion. Since the valve seat is parallel to the line of flow, globe valves are not recommended where resistance to flow and pressure drop are undesirable, because the design of the valve body is such that it changes the direction of flow causing turbulence and pressure drop within the valve. Globe valves have the highest pressure drop of any of the more commonly used valves. The shorter disc travel and the fewer turns to open and close a globe valve save time and wear on the valve stem and bonnet.

Angle valves, like globe valves, are used for throttling services. The flow on the inlet side is at right angles to the flow on the outlet side, making a 90-degree change in direction. The use of an angle valve eliminates the need for an elbow and extra fittings in the line. Basically the angle valve is a form of globe valve, having similar features of stem, disc, and seat ring.

Needle valves are a form of globe or angle valve in that the seating is similar. These valves are generally used as instrument valves. They derive their name from the needle point of the stem. Very accurate throttling can be handled by this type of valve. However, these valves are not normally recommended for steam service or where high temperatures will be encountered.

Service Recommendations

1. Designed for throttling service or flow regulation
2. Convenient for frequent operation — short stem travel of globe valves saves operator's time
3. Positive shut-off for gases and air
4. Applied where some resistance in line can be tolerated
5. Used where some fluid trapped in line is not objectionable.

Construction of Valve

Globe valves are available with a variety of discs and seats. The shapes of the disc and seat can be altered to provide different flow characteristics. However, there are only three basic types of discs:

1. Composition disc
2. Metal disc
3. Plug-type disc.

Composition Disc

The composition-type disc, shown in Fig. 4-6, has a flat face that is pressed against a flat, annular, metal seating surface. It consists of a metal disc holder, a composition disc of some suitable nonmetallic material, and a retaining nut. Among the materials used for composition discs are various rubber compounds, polytetrafluorethylene (Teflon,[1] Halon[2]), Kel-F,[3] and others, depending upon the valve manufacturer.

[1] Trademark, E. I. duPont.
[2] Trademark, Allied Chemical.
[3] Trademark, 3M Company.

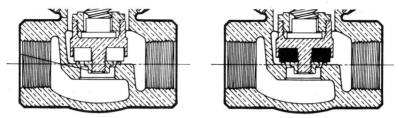

Courtesy of NIBCO Inc.

Fig. 4-6. Composition disc.

Each valve manufacturer recommends the type of composition disc which is best suited for various fluid media.

This type of disc is not recommended for throttling service but it does provide positive shut-off for gases and air. It also protects the valve seat from damage by dirt and scale, and the disc is quite easily and economically replaced.

Metal Disc

The metal disc, shown in Fig. 4-7, provides line contact against a conical seat with a tapered or spherical seating surface. When the proper materials are used for disc and seat, the line contact will break down deposits that may form on the seating surface. Soft materials, such as bronze, should only be used where the fluids are sediment free because they are easily damaged.

This type of disc is not recommended for throttling service but does provide positive shut-off for liquids.

Plug-Type Disc

The plug-type disc provides the best throttling service because of its configuration. It also offers maximum resistance to galling, wire draw-

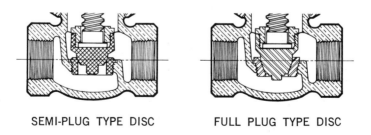

SEMI-PLUG TYPE DISC   FULL PLUG TYPE DISC

Courtesy of NIBCO Inc.

Fig. 4-7. Metal disc.

ing, and erosion. Plug-type discs are available in a variety of specific configurations, but in general they all have a relatively long tapered configuration. Each of the variations have specific types of applications and certain fundamental characteristics. Figure 4-8 illustrates the various types.

The equal percentage plug, as its name indicates, is used for equal percentage flow characteristic for predetermined valve performance. Equal increments of valve lift give equal percentage increases in flow.

Linear flow plugs are used for linear flow characteristics with high pressure drops.

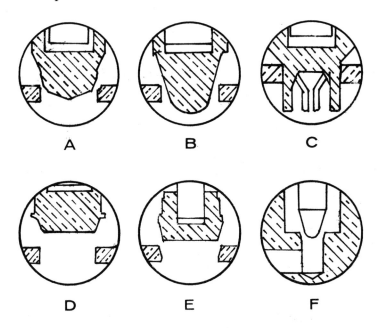

Fig. 4-8. Plug-type discs. A. Equal percentage plug; B. linear flow plug; C. "V" port plug; D. plug; E. semicone plug; F. needle plug.

"V" port plugs provide linear flow characteristics with medium and low pressure drops. This type of design also prevents wire drawing during low flow periods by restricting the flow, when the valve is only partially open, through the orifices in the "V" plug.

Needle plugs are used primarily for instrumentation applications and are seldom available in valves over 1 inch in size. These plugs provide high pressure drops and low flows. The threads on the stem are usually very fine; consequently, the opening between the disc and seat does not change rapidly with stem rise. This permits closer regulation of flow.

All of the plug configurations are available in either a conventional globe valve design or the angle valve design. Only when the needle plug is used is the valve name changed (needle valve). In all other cases the valves are still referred to as globe or angle valves with a specific type of disc.

Seat

Globe valve seats are either furnished cast integrally with the valve or furnished as seat rings which are screwed-in, pressed-in, or tack-welded in place. Seat rings with TFE inserts can also be supplied.

Integral seats provide a seat of the same material of construction as the valve body, while the other types permit a variation. This ability to vary the materials of construction of the seat from that of the valve body is advantageous in corrosive services.

The screwed-in seats are considered replaceable since they may be removed and new seat rings installed. This is an advantage in valves of expensive alloy construction in that it is not necessary to scrap the complete valve because of a damaged seat which cannot be repaired.

Stem Assemblies

Globe valves are available with stem assemblies of the following types:

1. Inside screw rising stem
2. Outside screw rising stem
3. Sliding stem.

For a discussion of the advantages, disadvantages, and recommended applications of each type of stem assembly, refer to pages 14 through 17.

Stem Seals

Various methods are available to prevent leakage along the valve stem to the outside of the valve. These are discussed starting on page 18.

Bonnet Construction

Globe valves come with a variety of bonnet constructions. Those generally available are:

1. Screwed-in bonnet
2. Screwed-on bonnet
3. Union bonnet
4. Flanged bonnet
5. Pressure sealed bonnet
6. Lip sealed bonnet
7. Breech lock bonnet.

A discussion of each of these types of bonnets will be found starting on page 22.

Identification of Parts

The nomenclature and identification of the different components and parts of globe valves with varying bonnet construction, stem assemblies, and types of discs are shown in Fig. 4-9, pages 72–74.

Available Materials of Construction

Globe valves are available in a wide range of materials of construction. Standard representative constructions include:

1. Bronze
2. All iron
3. Cast iron
4. Forged steel
5. Monel
6. Cast steel
7. Stainless steel.

Various trims are available on the above valves. Other metallic constructions are also available, some as standards and some on special order, depending upon the specific manufacturer.

Angle valves and needle valves can also be found in a variety of plastic materials, including:

1. Polyvinyl chloride (PVC)
2. Polypropylene
3. Penton
4. Impervious graphite.

Installation, Operation, and Maintenance

General installation procedures and care of valves prior to installation are given starting on page 40. Covered here are the items unique to globe valves.

Globe and angle valves should ordinarily be installed so that the pressure is under the disc. This promotes easy operation. It also helps to protect the packing and eliminates a certain amount of erosive action on the seat and disc faces. However, when high temperature steam is the medium being controlled, and the valve is closed with the pressure under the disc, the valve stem, which is now out of the fluid, contracts on cooling. This action tends to lift the disc off the seat, causing leaks which

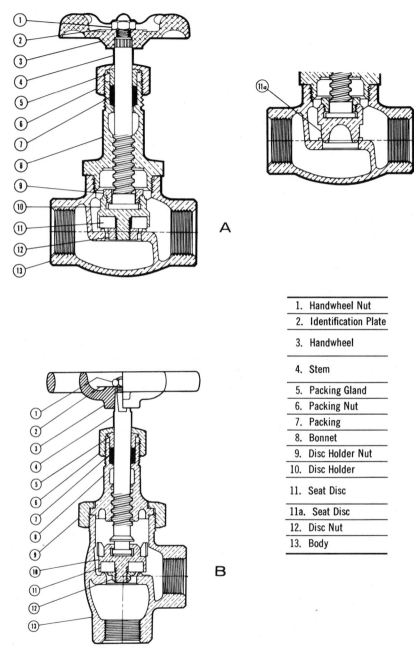

Fig. 4-9. Identification of globe valve parts. A. Screw-in bonnet composition disc; B. union bonnet composition disc angle body.

1. Handwheel Nut
2. Identification Plate
3. Handwheel
4. Stem
5. Packing Gland
6. Packing Nut
7. Packing
8. Bonnet
9. Disc Holder Nut
10. Disc Holder
11. Seat Disc
11a. Seat Disc
12. Disc Nut
13. Body

Courtesy of NIBCO Inc.

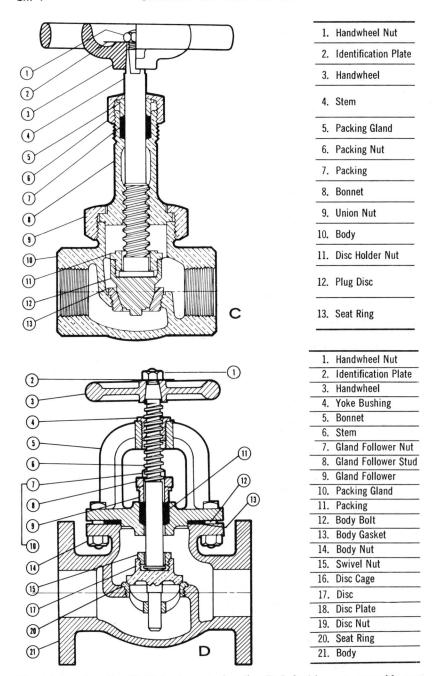

Fig. 4-9 (*continued*). C. Union bonnet plug disc; D. bolted bonnet renewable seat and disc.

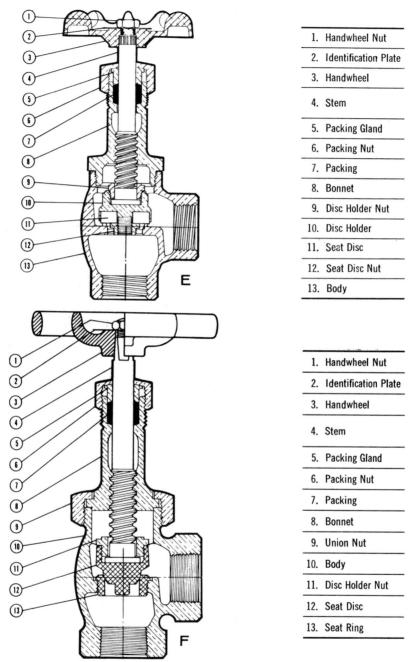

Fig. 4-9 (*continued*). E. Screw-in bonnet composition disc angle body; F. union bonnet semi-plug disc angle body.

eventually result in wire drawing on seat and disc faces. Therefore, in high-temperature steam service, globe valves should be installed so that the pressure is above the disc.

Lubrication of valves is especially important and should be done on a strict schedule. Valves that are opened and closed frequently should be lubricated at least once a month. On OS & Y valves where the stem is exposed, the screw threads should be kept clean and lubricated. Many valves are equipped with a lubrication fitting in the upper yoke to utilize pressure lubricant gun operation. Stem threads left dry and unprotected will become worn by grit and other abrasives, threatening stem failure.

Foreign matter on the seat of a globe valve can usually be flushed off the seat by opening the valve slightly, which creates a high rate of flow through the small opening provided.

If valves do not hold tight, do not use extra leverage, or wrenches on the handwheel, as a valve is easily ruined this way. Instead, take the valve apart and inspect the seat and disc to locate the trouble.

Packing leaks should be corrected immediately by tightening the packing nut which compresses the packing. If left unattended long enough, corrosive fluids will ruin the stem. This is true on both small and large valves. In addition, a leaking stem can lead to valve chatter in which the vibration will damage the working parts of the valve. If apparent that the packing gland has compressed the packing to its limit, replace with new packing, making sure that a packing is selected which is compatible with the materials being handled by the valve.

Regrindable seat valves should have a union-type bonnet construction for easy access, although bolted bonnet and screwed-in bonnet valves can also be reground. Remove the bonnet, by unscrewing the bonnet ring, or by removing the body bolts on a bolted bonnet. Place an ample amount of grinding compound on the disc, insert a pin in the groove of the disc holder and the hole in the stem, then reassemble bonnet to the body, screwing the union bonnet ring to hand tightness. Then back off one complete turn. Now the stem can be used as your regrinding tool. By reversing union bonnet ring only one complete turn, you assure yourself of the stem's being vertical and the disc and seat in perfect alignment. If the disc is off-center or cocked, the new reground seat will not be true. Do not overgrind, as unnecessary grinding on a seat and disc defeats the purpose of regrinding a renewable seat valve. When regrinding is completed, remove bonnet ring and bonnet, and thoroughly clean the regrinding compound from the seat and disc. Also, remove any scale or corrosive deposits which may have formed in the valve body or bonnet. Be sure to lubricate the threads before rejoining union bonnet ring and body for easy removal the next time.

Globe Valve Specifications

Once it has been decided that globe valves are to be used, a specification should be established that will provide a valve which will meet all of the requirements of the system. This specification should include the following items as well as the materials of construction of the different components.

1. Type of end connection
2. Type of disc
3. Type of seat
4. Type of stem assembly
5. Type of stem seal
6. Type of bonnet assembly
7. Pressure rating
8. Temperature rating.

4.3 Y-TYPE VALVES

Y-pattern valves are a modification of the globe valve but have features of the gate valve. They offer the straight way passage and unobstructed flow of the gate valve and the throttling ability of the globe valve. They have a lower pressure drop across the valve than the conventional globe valve. Such a valve is shown in Fig. 4-10.

All of the variations of disc and seat design available in the globe valve are also available in the Y-pattern valve. In general, any specifica-

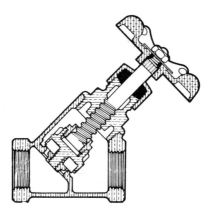

Courtesy of NIBCO Inc.

Fig. 4-10. Y-valve.

tion established for a conventional globe valve can be met by the Y-pattern valve except pressure drop, which is considerably lower in the Y-pattern valve.

These valves are also available in the same materials of construction as the conventional globe valves, plus borosilicate glass.

4.4 PLUG VALVES

General

Ancient ruins have uncovered wooden cocks, the forerunners of today's modern plug valves. The advantages of the plug valve are: minimum amount of installation space, simple operation, quick action (a 90-degree rotation of the plug opens or closes the fluid flow path), relatively little turbulence within the valve, and low pressure drop across the valve.

Plug valves (cocks) are normally used for nonthrottling on-off operations, particularly where frequent operation of the valve is necessary. These valves are not normally recommended for throttling service because, like the gate valve, a great percentage of flow change occurs near shut-off at high velocity. However, a diamond-shaped port has been developed for throttling service.

Another important characteristic of the plug valve is its easy adaption to multiport construction. Multiport valves are widely used. Their installation simplifies piping, and they provide a much more convenient operation than multiple gate valves would. They also eliminate pipe fittings. The use of a multiport valve eliminates the need of two, three, or even four conventional shut-off valves, depending upon the number of ports in the plug valve.

Plug valves are available in either a lubricated or nonlubricated design and with a variety of styles of port openings through the plug as well as a number of plug designs.

Service Recommendations

1. Fully opened or fully closed nonthrottling service, except for valves fitted with special plug ports to allow throttling
2. Minimum resistance to flow
3. Minimum amount of fluid trapped in line
4. Frequent operation
5. Low pressure drop
6. Use as a flow-diverting valve.

Construction of Valve

The basic components of the plug valve are the body, plug, and gland. In addition to those with a conventional straight-through body, these valves are also used as:

1. Short pattern valves with compact face-to-face dimensions conforming to ANSI B16.10 (standard for gate valves). Port openings can be either full or reduced but are usually rectangular in shape.

2. Multiport valves which have bodies with three or more pipe connections and are used for transfer or diverting services.

3. Venturi design which is a reduced area, round or rectangular port with Venturi flow through the body. Usually the minimum port opening is 40 percent.

Multiport Valves

This body style is particularly advantageous on transfer lines and for diverting services. A single multiport valve may be installed in lieu of three or four gate valves, or other type of shut-off valves. One word of caution is that many of the multiport configurations do not permit complete shut-off of flow. In most cases one flow path is always open. These valves are intended to divert the flow to one line while shutting off flow from the other lines. If complete shut-off of flow is a requirement, it is necessary that a style of multiport valve be used which permits this, or a secondary valve should be installed on the main feed line ahead of the multiport valve to permit complete shut-off of flow.

It should also be noted that in some multiport configurations, flow to more than one port simultaneously is also possible. Great care should be taken in specifying the particular port arrangement required to guarantee that proper operation will be possible.

Figure 4-11 shows typical port arrangements on multiport valves.

Plug

Plugs are either round or cylindrical with a taper. They may have various types of port openings, each with a varying degree of free area relative to the corresponding pipe size. Port types are as follows:

1. Rectangular port is the standard shaped port with a minimum of 70 percent of the area of the corresponding size of standard pipe.

2. Round port means that the valve has a full round opening through the plug and body, of the same size and area as standard pipe.

3. Full port means that the area through the valve is equal to or greater than the area of standard pipe.

Ch. 4 GENERAL PURPOSE VALVES

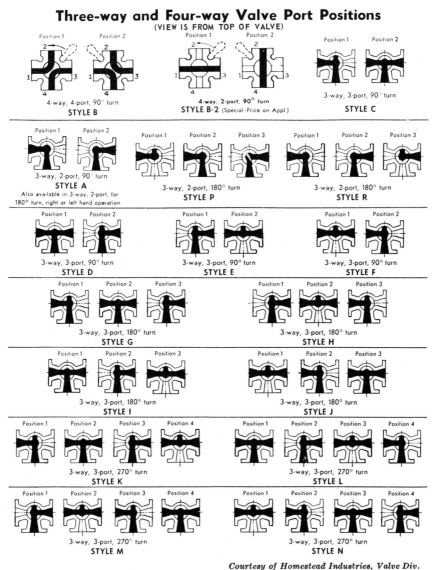

Fig. 4-11. Typical multiport arrangements.

4. Standard opening means that the area through the valve is less than the area of standard pipe, and therefore these valves should be used only where restriction of flow is unimportant.

5. Diamond port means that the opening through the plug is diamond-

ROUND PORT PLUG   **DIAMOND PORT PLUG**   **RECTANGULAR PLUG (Standard)**

Courtesy of Homestead Industries, Valve Div.

Fig. 4-12. Typical plug openings.

shaped. This has been designed for throttling service. All diamond port valves are venturi restricted flow type.

Figure 4-12 shows typical plug openings.

Lubricated Plug Valves

Lubricated plug valves have certain advantages over other conventional types of valves. These valves have the ability to give tight shut-off on hard-to-hold substances. A quarter turn of the plug allows the valves to be opened or closed. The plug is the only moving part of the valve when in operation, and its tapered configuration assures positive seating.

The plug is designed with grooves which permit a lubricant to seal and lubricate the valve as well as to function as a hydraulic jacking force to lift the plug within the body, thus permitting easy operation. The straight-way passage through the port offers no opportunity for sediment or scale to collect. The valve plug, when rotated, wipes foreign matter from the plug. However, these valves must be kept lubricated at all times to maintain a tight seal between the plug and body.

Lubricated plug valves have these distinct characteristics:

1. Tight shut-off if proper lubrication is used
2. Quick operation
3. Shearing action of plug, which wipes off scale and foreign matter
4. Lubrication necessary to prevent galling and sticking
5. Not recommended for low or high temperature service.

The correct choice of lubricant is extremely important for successful lubricated plug valve performance. In addition to providing adequate lubrication to the valve, the lubricant must not react chemically with the material passing through the valve nor must the lubricant contaminate the material passing through the valve because of solubility. All manufacturers of lubricated plug valves have developed a series of lubricants which are compatible with a wide range of media. Their recommendation should be followed as to which lubricant is best suited for the service.

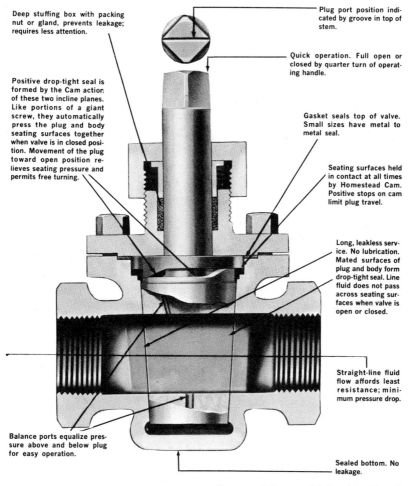

Courtesy of Homestead Industries, Valve Div.

Fig. 4-13. Cam-operated nonlubricated plug valve.

Nonlubricated Plug Valves

In order to overcome the disadvantages of the lubricated plug valves, in so far as their need of lubrication, two basic types of nonlubricated plug valves were developed. The nonlubricated valve may be either a lift-type or have an elastomer sleeve or plug coating that eliminates the need to lubricate the space between the plug and seat.

Lift-type valves provide a means of mechanically lifting the tapered plug slightly to disengage it from its seating surface to permit easy rotation. The mechanical lifting can be accomplished through either a cam, as shown in Fig. 4-13, or by means of an external lever as shown in Fig. 4-14.

A typical nonlubricated plug valve with an elastomer sleeve is shown in Fig. 4-15. In this particular valve a sleeve of TFE completely surrounds the plug. It is retained and locked in place by the metal body. This results in a continuous primary seal being maintained between the sleeve and the plug at all times, both while the plug is rotated and when the valve is in either the open or closed position. The TFE sleeve is durable and essentially inert to all but a few rarely encountered chemicals. It also has a low coefficient of friction and therefore is self-lubricating.

Nonlubricated plug valves have the following characteristics:

1. Absolute shut-off — can use resilient seats
2. Quick operation
3. Suitable for wider temperature ranges than lubricated plug valves
4. Products handled not subject to contamination by a lubricant
5. Best for handling slurries — will not stick
6. Not necessary to lubricate valve — low maintenance cost.

Gland

The gland of the plug valve is equivalent to the bonnet of a gate or globe valve. It is that portion of the valve assembly which secures the stem assembly to the valve body. There are three general types of glands — single gland, screwed gland, and bolted gland.

Identification of Parts

The nomenclature and identification of the different components and parts of plug valves are shown in Fig. 4-16.

Available Materials of Construction

Plug valves are available in a wide range of materials of construction including solid plastics and many lined configurations. Among the more

Ch. 4   GENERAL PURPOSE VALVES   83

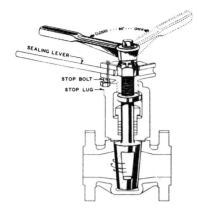

Position of sealing — lever stop bolt shows whether sealing pressure has, or has not, been relieved. Sealing pressure has been relieved when stop bolt rests against lug. Do not attempt to operate valve if sealing-lever stop bolt is in any other position. Sealing pressures must be restored after each operation.

Features

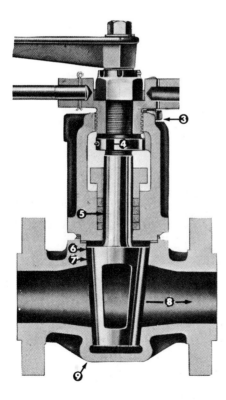

❶ Plug lever. Quarter-turn to open or close.

❷ Sealing lever. Valve cannot stick. The powerful lever and screw principle assures positive operation. It is part of the valve and always ready for instant use.

❸ Positive stop allows seating pressure to be relieved only enough to overcome friction between plug and body and permit easy turning of the plug.

❹ Visible outside stop limits plug travel to quarter-turn, and assures alignment of plug and body ports. Plug lever lines up with groove on top of stem to indicate port openings.

❺ Deep stuffing box and gland. No leakage of valuable fluids.

❻ "Lever-seald" action presses mated metal surfaces firmly together to form perfect seal against leakage.

❼ Protected seating surfaces. No fluid or grit passes across seating surfaces in either open or closed position. Maximum valve life and lowest cost-per-year service.

❽ Straight-line fluid flow. Low pressure drop.

❾ Sealed bottom prevents leakage.

Courtesy of Homestead Industries, Valve Div.

Fig. 4-14.   Lever-operated nonlubricated plug valve.

Courtesy of Xomox Corp.

Fig. 4-15. Plug valve with elastomeric sleeve.

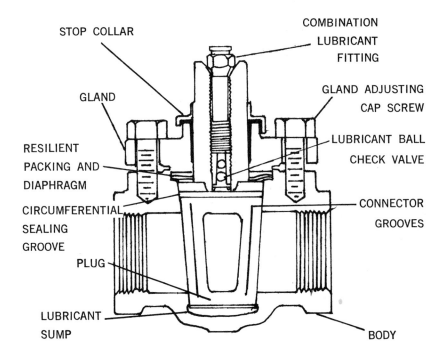

Fig. 4-16. Identification of plug valve parts.

readily available materials of construction are:

Bronze	Polyvinylchloride (PVC)
Steel	TFE-lined
Stainless steel	Ni-resist
Brass	Penton

Installation, Operation, and Maintenance

General installation procedures and care of valves prior to installation are given starting on page 40. Covered here are the specific items unique to plug valves.

When installing plug valves, care should be taken to allow room for the operation of the handle (lever, wrench). The handle is usually longer than the valve, and it rotates to a position parallel to the pipe from a position 90 degrees to the pipe.

Lubricants are available in stick form, in tubes, and in bulk. Stick lubrication is usually employed when a small number of valves are in service or when they are widely scattered throughout the plant. However, for a large number of valves gun lubrication is the most convenient and economical solution. Tube lubrication is used as stick lubrication. Not all types of lubricants are made in stick form.

All valves are usually shipped with an assembly lubricant. This assembly lubricant should be removed by completely relubricating with the proper lubricant before being put into service.

Gun Lubrication

1. Valve should be in fully open or fully closed position. The best results are obtained with the valve in the fully open position.
2. Connect high-pressure lubricant gun to combination lubricant fitting and pump lubricant into valve until resistance is felt.
3. Rotate plug back and forth during final strokes of lubricant gun to spread lubricant evenly over the plug and body seating surfaces.

Stick Lubrication

1. Valve should be in the fully open or fully closed position. The best results are obtained with the valve in the fully open position.
2. Remove lubricant fitting and insert proper size stick of lubricant.
3. Replace fitting. Lubricant will be forced into the valve by turning the fitting.
4. Repeat, adding additional sticks of lubricant until increased resistance is felt on fitting, which indicates that the lubricant system is full and under pressure.
5. Rotate plug back and forth while turning in last stick of lubricant to spread the lubricant evenly over the plug and body seating surfaces.

Regular periodic lubrication is a must for best results. Extreme care should be taken to prevent any foreign matter from entering the plug when inserting new lubricant.

Gland Adjustment

To insure a tight valve the plug must be seated at all times. Gland adjustment should be kept tight enough to prevent the plug from becoming unseated and exposing the seating surfaces to the live fluid. Care should also be exercised so as not to overtighten the gland, which will result in a metal-to-metal contact between the body and the plug. Such a metal-to-metal contact creates an additional force which will require extreme effort to overcome when operating the valve.

Hard Operation

1. Lubricate valve thoroughly with the proper lubricant. Hard operation can sometimes result from using the improper lubricant.
2. Should valve remain difficult to operate after lubrication, loosen gland (or bonnet) slightly and relubricate valve thoroughly.
3. If gland (or bonnet) has been loosened to the extent that the valve operates freely, retighten gland (or bonnet) until the proper operation is obtained. Freely operating valves require an ample amount of lubricant to keep them from leaking.
4. If the above operations do not free the plug, it will be necessary to disassemble the valve and clean it thoroughly. Be sure that all hardened lubricant is removed from the lubricant system. Reassemble and lubricate thoroughly.

Leakage

1. Lubricate valve thoroughly.
2. If grease backpressure fails to build up while lubricating, the valve is out of adjustment. The plug is not seated in the body and the lubricant that is added will be dispersed into the line.
3. Tighten gland (or bonnet) while operating the plug back and forth to work out the excess lubricant. Gland (or bonnet) should be tightened so that valve operates with moderate effort.

Plug Valve Specification

A specification for plug valves should include, besides basic materials of construction, the following items:

1. Type of plug
2. Type of gland

3. Lubricant to be used (if lubricated plug valve)
4. Port arrangements if multiport design
5. Temperature rating
6. Pressure rating.

4.5 BALL VALVES

General

The ball valve, in general, is the least expensive of any valve configuration. In early designs having metal-to-metal seating the valves could not give bubble-tight sealing and were not firesafe. With the development of elastomeric materials and with advances in the plastics industries, the original metallic seats have been replaced with materials such as fluorinated polymers, nylon, neoprene, and buna-N.

The ball valve is basically an adaptation of the plug valve. Instead of a plug it uses a ball with a hole through one axis to connect the inlet and outlet ports in the body. The ball rotates between resilient seats. In the open position, the flow is straight through. When the ball is rotated 90 degrees the flow passage is completely blocked.

In addition to quick, quarter-turn on-off operation, ball valves are compact, easy-to-maintain, require no lubrication, and give tight sealing with low torque.

Since conventional ball valves have relatively poor throttling characteristics, they are not generally satisfactory for this service. In a throttling position the partially exposed seat rapidly erodes because of the impingement of high-velocity flow. However, a ball valve has been developed with a spherical surface-coated plug which is off to one side in the open position and which rotates into the flow passage until it blocks the flow path completely. Seating is accomplished by the eccentric movement of the plug. The valve requires no lubrication and can be used for throttling service.

Service Recommendations

1. Designed primarily for on-off service
2. For minimum resistance to flow
3. Quick opening
4. Low maintenance cost
5. Limited to moderate temperature services.

Construction of Valve

Ball valves are designed on the simple principle of floating a polished ball between two plastic O-ring seats, permitting free turning of the ball.

Since the plastic O-rings are subject to deformation under load, some means must be provided to hold the ball against at least one seat. Normally this is accomplished through spring pressure, differential line pressure, or a combination of both. In essence, ball valves seal on the same theory as a check valve.

With a soft seat on both sides of the ball, most ball valves give equally effective sealing of flow in either direction. Most designs permit adjustment for wear.

Ball valves are available in the venturi, reduced, and full port patterns. The latter has a ball with a bore equal to the inside diameter of the pipe. Balls are usually metallic in metallic bodies with trim (seats) produced from elastomeric materials. Also, all-plastic designs are available. Seats are replaceable, as are the balls.

Ball valves are available in top entry and split body (end entry) types. In the former, the ball and seats are inserted through the top, while in the latter, the ball and seats are inserted from the ends as shown in Fig. 4-17.

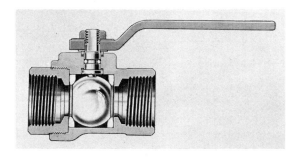

Courtesy of Lunkenheimer

Fig. 4-17.   End entry ball valve.

Another design of ball valve provides for the interchangeability of the end pieces in various materials of construction. This permits the changing of the end pieces, if necessary, as long as galvanic action will not present a problem.

Ball valves are also available in multiport configurations in the same manner as plug valves. The typical port arrangements are the same as shown in Fig. 4-11 for plug valves. These valve bodies are usually larger than the equivalent straight-way valve since a larger diameter ball must be provided to permit the multiple port drillings.

When closed, many ball valves trap the fluid between the seats and in the hole in the ball, which may be objectionable in some cases. This type of valve is limited to temperatures and pressures allowed by the seat material.

Seats

The resilient seats for ball valves are made from various elastomeric materials. The most common seat materials are TFE (virgin), filled TFE, Nylon, Buna-N, Neoprene, and combinations of the foregoing. Because of the elastomeric materials these valves cannot be used at elevated temperatures. Typical average *maximum operating* temperatures at full pressure rating of the valves are shown in Table 4-3 for various seat materials. In order to overcome this disadvantage a graphite seat has been developed which will permit operation up to 1000°F.

Table 4-3. Maximum Operating Temperature of Ball Valve Seats

Seat Material	TFE (Virgin)	Filled TFE	Buna-N	Neoprene
Oper. Temp. °F*	280	325	180	180

* These are average values and will vary between valve manufacturers.

Care must be used in the selection of the seat material to insure that it is compatible with the materials being handled by the valve, as well as its temperature limitation.

Stem and Bonnets

The stem in a ball valve is not fastened to the ball. It normally has a rectangular portion at the ball end which fits into a slot cut into the ball. This engagement permits rotation of the ball as the stem is turned.

A bonnet cap fastens to the body which holds the stem assembly and ball in place. Adjustment of the bonnet cap permits compression on the packing which supplies the stem seal. Packing for ball valve stems is usually in the configuration of die-formed packing rings normally of TFE, TFE-filled, or TFE-impregnated materials. Some ball valve stems are sealed by means of O-rings rather than packing.

Some ball valve stems are equipped with stops which permit only a 90-degree rotation, while others do not have stops and may be rotated 360 degrees. In the latter valves a 90-degree rotation is still all that is required for closing or opening of valve. Position of the handle indicates valve ball position. When the handle lies along the axis of the valve, the valve is open; when the handle lies 90 degrees across the axis of the valve, the valve is in the closed position.

In addition to the handle position as an indicator of valve setting, some ball valve stems have a groove cut in the top face of the stem which shows the flow path through the ball. Observation of the groove position indicates the position of the port through the ball. This feature is particularly advantageous on multiport valves. Figure 4-18 shows two such stem grooves, one for a straight through valve and one for a multiport valve.

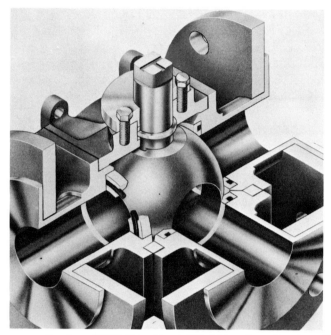

Courtesy of WKM Valve Div. of ACF & Dover Corp., Ronningen-Petter Div.

Fig. 4-18. Grooves in ball valve stem.

Identification of Parts

The nomenclature and identification of the components and parts of ball valves are shown in Fig. 4-19.

Available Materials of Construction

Ball valves are available in an extremely wide range of materials of construction including:

cast iron	aluminum	borosilicate glass
ductile iron	carbon steel	impervious graphite
bronze	stainless steels	polypropylene
brass	titanium	polyvinylchloride
zirconium	tantalum	Penton

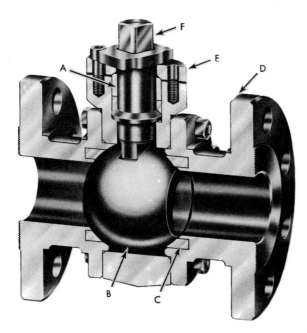

Courtesy of WKM Valve Div., ACF Industries

Fig. 4-19. Identification of ball valve parts. A. Packing, B. ball, C. seat, D. body, E. bonnet cap, F. stem.

Installation, Operation, and Maintenance

General installation procedures and care of valves prior to installation are given starting on page 40. Covered here are the specific items unique to ball valves.

When installing ball valves, care should be taken to allow sufficient room for the operation of the valve handle. The valve handle is usually larger than the valve body, and it rotates from a position parallel to the pipe to a position 90 degrees across the pipe. Conventional ball valves should not be used for throttling service.

Weld End Valves

To prevent damage to the ball and seats due to excessive heat or weld slag, the following procedure is recommended for welding ball valves into pipelines:

1. Place the ball in the fully open position before starting to weld.
2. Use an internal welding back-up ring where practical.

3. Avoid rapid application of excess welding material.

4. Do not permit the temperature of valve body-seat area to exceed 250°F, to prevent seat and seal damage. (Check with Tempil Stick.)

5. Before turning valve to the closed position clean the pipeline and valve bore of weld slag. This can be done by pigging and flushing the line.

6. Alternative to this procedure is to remove the ball and seats, weld the valve in place, clean the pipeline of weld slag, and reassemble the ball and seats.

Socket Weld-End Valves

In addition to the foregoing standard welding procedure, the following special procedure should also be followed for socket weld-end valves:

1. Provide adequate support to the pipe on each side of the valve prior to welding.

2. Use an electric welding rod (maximum ⅛-inch diameter). Weld each end of valve with a continuous bead. Do not apply excessive weld metal.

3. If multiple passes are necessary, make certain that the interpass temperature does not exceed 250°F in the body seat area. (Check with Tempil stick.)

If it is necessary to disassemble a ball valve in order to replace one or more portions the following procedure should be adhered to on the reassembly:

1. Ball should be in the closed position.

2. If the top of the stem is slotted, the slot should line up with the port in the ball.

3. Inspect the seats and make sure that they are not damaged or scored.

4. Torque up the bolts or cap screws evenly, alternating back and forth across the valve. Also use a feeler gage if necessary to obtain a uniform gap between end fitting and body.

5. Rotate ball back and forth while torquing up on bolts or cap screws, to avoid getting it too tight.

Packing leaks should be taken care of immediately by tightening the gland which compresses the packing. The gland must be tightened uniformly alternating back and forth across the valve. Do not overtighten. Rotate the ball back and forth while tightening to prevent getting it too tight. If it is necessary to tighten excessively, it is an indication that the packing should be replaced. Overtightening can damage and deform the elastomeric seats.

Ball Valve Specifications

A ball valve specification should include, besides the basic materials of construction of the body, the following items:

1. Operating temperature
2. Operating pressure
3. Type of port in ball and whether full port or reduced port is required.
4. Material of construction of seat
5. Whether ball is to be top entry or end entry.

4.6 BUTTERFLY VALVES

General

Butterfly valves possess many advantages over gate, globe, plug, and ball valves in a variety of installations, particularly in the larger sizes. Savings in weight, space, and cost are the most obvious advantages. Maintenance costs are low since there is a minimum number of moving parts and there are no pockets to trap fluids.

This type of valve is suitable for throttling as well as open-closed applications. Operation is easy and quick because a 90-degree rotation of the handle moves the flow control element from the fully closed to the fully opened position. These valves may also be equipped for automatic operation.

Butterfly valves are especially well suited for the handling of large flows of liquids or gases at relatively low pressures and for the handling of slurries or liquids with large amounts of suspended solids.

Service Recommendations

1. Fully opened, fully closed, or throttling service
2. Low pressure drop across the valve
3. Minimum amount of fluid trapped in the line
4. Frequent operation
5. Positive shut-off for gases or liquids
6. Handling of slurries.

Construction of Valve

Butterfly valves are built on the principle of a pipe damper. The flow control element is a disc of approximately the same diameter as the inside diameter of the adjoining pipe, which rotates on either a vertical or horizontal axis. When the disc lies horizontally, the valve is fully opened, and when the disc approaches the vertical position, the valve is shut. Intermediate positions, for throttling purposes, can be secured in place by handle-locking devices.

Stoppage of flow is accomplished by the valve disc sealing against a seat which is on the inside diameter periphery of the valve body. Originally a metal disc was used to seal against a metal seat. This arrangement did not provide a leak-tight closure but did provide sufficient closure in such applications as water distribution lines. Valves of this design are still available.

With the advent of the newer elastomeric materials most butterfly valves are now produced with an elastomeric seat against which the disc seals. This arrangement does provide a leak-tight closure.

Body construction varies. The most economical is the wafer type which simply fits between two pipeline flanges. Another type of lug wafer valve is held in place between two pipe flanges by bolts that join the two flanges and pass through holes in the valve's outer casing. Valves are also available with conventional flanged ends for bolting to pipe flanges and in a screwed end construction. These various configurations are shown in Fig. 4-20. Butterfly valves are also available in sanitary construction.

Disc

Discs for butterfly valves are available in a wide range of materials of construction, including practically all metals. Metal discs are also available with many types of coverings such as TFE, Kynar, Buna-N, Neoprene, Hypalon, and other corrosion-resistant elastomers. Discs are replaceable.

Seat

Butterfly valve seats can be of metal or resilient liner types. The former do not give a leak-proof closure, while the resilient liner types do provide a leak-proof closure. In most valves the resilient liner is an elastomeric material which is replaceable. Typical materials used as seats are as follows:

Buna-N	Black Neoprene	Hypalon
Viton	Natural Rubber	Hycar
EPT Nordel White	White Butyl	TFE
EPT Nordel Black	Polyurethane	

TWO-FLANGE BODY

LUG WAFER BODY

SCREWED END VALVE WITH LEVER AND
INDICATION DIAL PLATE

WAFER BODY

Courtesy of Crane Co. (upper left and right, lower right)
Courtesy of Rockwell Mfg. (lower left)

Fig. 4-20. Types of butterfly valve bodies.

Stem and Gland Assembly

The stem and disc are separate pieces. The disc is bored to receive the stem. Two methods are used to secure the disc to the stem so that the disc rotates as the stem is turned.

In the first method the disc is simply bored through and the disc secured to the stem by means of bolts or pins which pass through the stem and disc as shown in Fig. 4-21. The alternate method, also shown

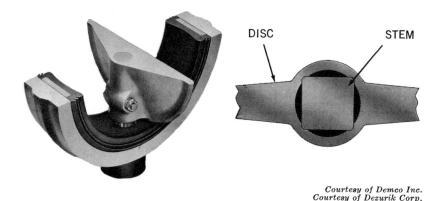

Courtesy of Demco Inc.
Courtesy of Dezurik Corp.

Fig. 4-21. Connection of stem to disc.

in Fig. 4-21, involves boring the disc as before, then broaching the upper stem bore to fit a squared stem. This method allows the disc to "float" and seek its center in the seat. Uniform sealing is accomplished and external stem fasteners are eliminated. This is advantageous in the case of covered discs and in corrosive applications.

In order for the disc to be held in the proper position, the stem must extend beyond the bottom of the disc and be fitted into a bushing in the bottom of the valve body. One or two similar bushings are also necessary along the upper portion of the stem as well. These bushings must either be resistant to the media being handled or they must be sealed so that the corrosive media cannot come into contact with them. Both methods are employed, depending upon the valve manufacturer.

Stem seals are effected either with packing in a conventional stuffing box or by means of O-ring seals. Some valve manufacturers, particularly those specializing in the handling of corrosive materials, effect a stem seal on the inside of the valve so that no material being handled by the valve can come into contact with the valve stem. If a stuffing box or external O-ring seal is employed, the material passing through the

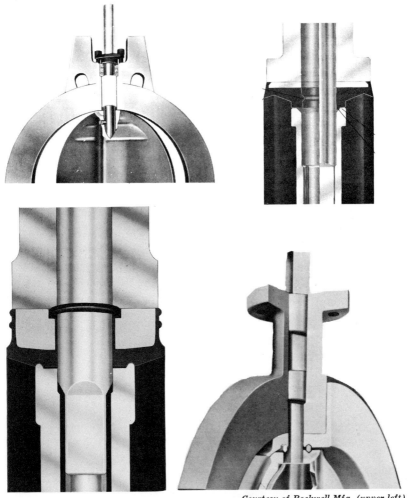

Courtesy of Rockwell Mfg. (upper left)
Courtesy of Demco Incorporated (upper right, lower left)
Courtesy of Garlock Inc. (lower right)

Fig. 4-22. Stem-sealing arrangements.

valve will come into contact with the valve stem. Different types of sealing arrangements are shown in Fig. 4-22.

Identification of Parts

The nomenclature and identification of parts of butterfly valves are shown in Fig. 4-23.

98 GENERAL PURPOSE VALVES Ch. 4

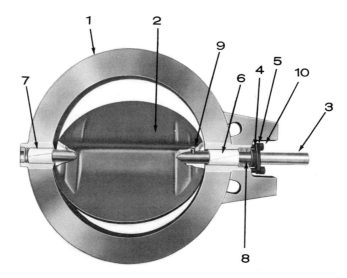

1—Body
2—Disc
3—Shaft
4—Gland
5—Gland retainer
6—Upper bearing
7—Lower bearing
8—Packing
9—Pin
10—Gland screws

Courtesy of Rockwell Mfg.

Fig. 4-23. Identification of butterfly valve parts.

Courtesy of Continental Div. Fisher Controls Co.

Fig. 4-24. Lever-actuated butterfly valve.

Ch. 4   GENERAL PURPOSE VALVES   99

Installation, Operation, and Maintenance

General installation procedures and care of valves prior to installation are given starting on page 40. Covered here are the specific items unique to butterfly valves.

Butterfly valves can be operated manually by lever (handle), handwheel, or chainwheel. Figure 4-24 shows a lever actuator. A steel or ductile iron handle with a spring-loaded locking trigger is secured to the stem. The position indicator has increments for variable control between 0 and 90 degrees nonadjustable travel stops. Usually lever actuators are not recommended on valves above 10 or 12 inches in size because of the high torque required for operation.

A handwheel actuator is shown in Fig. 4-25. A totally enclosed screw and lever mechanism imparts handwheel movement to the butterfly valve disc for variable settings between full open and full closed positions. The disc position indicator consists of shaft mounted pointer and cast-in increment scale on housing cover plate. This type of actuation is recommended for high torque conditions and corrosive atmospheres. It should be noted that this type of actuator is self locking in any position. Extension handles and floor stand handwheels may also be used on these handwheel actuators.

Courtesy of Continental Div. Fisher Controls Co.

Fig. 4-25.   Handwheel-actuated butterfly valve.

Courtesy of Continental Div. Fisher Controls Co.

Fig. 4-26. Chainwheel-actuated butterfly valve.

Chainwheel actuators are modifications of the handwheel actuators in which the chainwheel operator has been incorporated into the assembly rather than using the handwheel itself. Such a unit is shown in Fig. 4-26.

In addition to manual operation, butterfly valves may also be operated by means of air or fluid power and electricity. Available for this type of operation are cylinder actuators, diaphragm actuators, piston P.O.P. actuators, and electric actuators. These are shown in Fig. 4-27.

Operating torques required for specific valves and services should always be checked with the valve manufacturer before sizing the automatic operating accessory. Most manufacturers will supply the valve with the automatic operation if specified.

Care should be taken in planning the piping layout to allow sufficient room for operation of the handle (if lever-actuated).

Most butterfly valves are designed for installation between ANSI 150-pound flanges in schedule-40 piping systems. All flange types are permissible provided the inside diameter of the flange at the face does not differ significantly from the pipe inside diameter. Wall sections heavier or lighter than schedule 40 may require spacers in order to provide adequate disc clearance or support for the valve seat and liner.

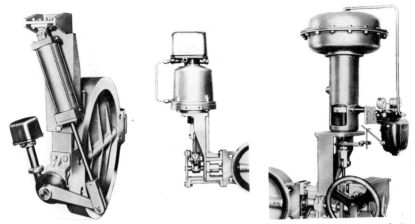

Courtesy of Continental Div. Fisher Controls Co.

Fig. 4-27. Automatic actuators for butterfly valves.

Special care should be taken in checking these dimensions when a lined piping system is being used. Valve manufacturers can supply the dimensional limitations on pipe dimensions and the proper spacer dimensions. Spacers, when required, must be installed on both sides of the valve.

Gaskets are not normally required on valves having resilient seats since the elastomeric material which forms the valve liner usually forms a gasket on both flange faces. Gaskets may be used, and are recommended, for protection of the liner, where frequent disassembly of the associated piping may be required. Gaskets should be used when the valve is to be installed between smooth face (ground or rigid plastic) or glass-lined pipe flanges.

Valves should remain in the closed position during all handling and installation operations. This is necessary to protect the disc edge or sealing surface from nicks and scratches. Such damage would impair the bubble-tight seal of the valve.

Care should also be taken to protect the valve liner from damage during handling. Severe scratches or damage to the liner may not be overcome by the flange pressure, making a gasket seal impossible. When slipping the valve between flanges, it is very important that the liner not be allowed to catch on the pipe I.D. and fold over. This will cause flange leakage and severe damage to the liner.

When tightening flange bolts, normal wrench torques may be used without fear of damage to the valve or liner.

After the valve has been installed between flanges and all flange bolts have been tightened, slowly turn the disc and check for freedom of disc movement.

If the valve is to be removed from the pipe line for any reason, the valve should be closed before the flange bolts are loosened, and remain closed until removed from the pipeline. Do not run sharp instruments between the valve and the liner or between the liner and the pipe flanges. This practice will result in severe liner damage.

Adjustment

Adjustment of the valve disc to engage the seal properly is important for bubble tight shutoff. The means of making this adjustment varies between valve manufacturers and should be checked before an attempt is made to make any adjustment.

Normally butterfly valves do not require any maintenance other than checking to see that there is no stem leakage. If there is stem leakage it should be corrected as soon as possible. The steps to take will vary depending upon the style of stem seal and the manufacturer. Instructions should be followed which are supplied by the manufacturer.

Butterfly Valve Specifications

Specifications established for butterfly valves should include the following items:

1. Type of body
   A. Wafer
   B. Lug wafer
   C. Flanged
   D. Screwed
2. Type of seat
   A. Metal
   B. Resilient
3. Material of construction
   A. Body
   B. Disc
   C. Seat
4. Type of actuation
5. Operating pressure
6. Operating temperature

4.7 DIAPHRAGM VALVES

Diaphragm valves are particularly suited for the handling of corrosive fluids, sticky and/or viscous materials, fibrous slurries, sludges, foods, pharmaceuticals or other products which require high purity and must remain free from contamination. The operating mechanism of a diaphragm valve is not exposed to the media within the pipeline. Sticky or viscous fluids cannot get into the bonnet to interfere with the operating mechanism. Many fluids that would clog, corrode, or gum up the working parts of most other types of valves will pass through a diaphragm valve without causing problems. Conversely, lubricants used for the operating mechanism cannot contaminate the fluid being handled. There are no packing glands to maintain and there is no possibility of stem leakage.

Diaphragm valves can be used for on-off service and throttling service. There are two general body types: the weir or Saunders patent, which is the most widely used, and the straight through type. The weir type is the better throttling valve but has a limited range. Its throttling characteristics are essentially those of a quick-opening valve because of the large shut-off area along the seat. Thus the lower part of the throttling curve is useless for most control purposes. However, there is a weir-type diaphragm valve that is capable of controlling small flows. Key to the improved range is a two-piece compressor component. (See page 109 for identification of parts.) Instead of the entire diaphragm lifting off the weir when the valve is opened, the first increments of stem travel raise an inner compressor component that causes only the central part of the diaphragm to lift. This creates a relatively small opening through the center of the valve. After the inner compressor is opened to its limit, the outer compressor component is raised along with the inner one and the remainder of the throttling is similar to that which takes place in the conventional valve.

Service Recommendations

1. Fully opened, fully closed, or throttling service
2. Handling of slurries, highly corrosive materials, or materials to be protected from contamination
3. Service with low operating pressures.

Construction of Valve

Diaphragm valves are, in effect, simple "pinch clamp" valves. A resilient, flexible diaphragm is connected to a compressor by a stud molded into the diaphragm. The compressor is moved up and down by the valve stem. Thus the diaphragm is lifted high when the compressor is raised. As the compressor is lowered, the diaphragm is pressed tight against the body weir (in the weir-type valve) or the contoured bottom in the straight-through valve. There is another style of valve in which a plug and diaphragm are molded into a single unit. In the open position, the valve provides straight-line flow, but the cross sectional area is somewhat less than that of the pipe. The body is a venturi pattern. When closed, the slightly tapered plug seats on the bottom of the body and into a port at the top of the flow passage through which the plug passes down. There is no line pressure on the diaphragm in the closed position.

The major components of the diaphragm valve are the body, the diaphragm, and the stem and bonnet assembly. There are no seats as such in a diaphragm valve since the body itself acts as the seat.

Body

There are two general styles of bodies — weir-type and straight-through type. Of the two the weir type is the most widely used, primarily because of the wider range of diaphragm materials available.

Bodies are available in an extremely wide range of materials of construction, including lined constructions. Table 4-4 shows the materials in which the bodies of diaphragm valves may be had.

Table 4-4. Body Materials for Diaphragm Valves

Metallic Bodies		
Cast iron	Alloy 20	Everdur
Cast steel	Bronze	Aluminum
Stainless steel	Ductile iron	Leaded red brass
Hastelloy A, B, C	Monel	Titanium
Solid Plastic Bodies		
Polyvinylchloride (PVC)	Polyvinylidene chloride (Saran)	Chlorinated polyether (Penton)
Polypropylene	Blue asbestos reinforced epoxy (Chemtite EB)	Blue asbestos reinforced phenolic (Chemtite PB)
Vinylidene fluoride	ABS	
Lined Bodies		
Hard natural rubber	Soft natural rubber	Chlorinated polyether (Penton)
Glass	Lead	Heresite
Polyvinyl chloride (PVC)	Polyvinylidene chloride (Saran)	Polypropylene
Lithcote	Porcelain	Fluorinated ethylene propylene (FEP)
Vinylidene fluoride (Kynar)	Titanium	Ethylene propylene (EPT)
Butyl	Buna-N	Neoprene

All of the body materials shown are not necessarily available in all types of end configurations. For example, the lined valves are only available with flanged ends, except for those of hard rubber, soft rubber, neoprene, and glass which are available on valves with Victaulic[1] ends. Most of the solid plastic and metallic valves are available with either screwed or flanged ends. Metallic valves are also available with weld ends, sanitary threads, and Victaulic ends. Bodies are also available in angle design.

Diaphragms

Just as there is a wide choice of body materials there is also a wide choice of available diaphragm materials. Diaphragm life depends not only upon the nature of the material handled but also upon the temperature, pressure, and frequency of operation. Diaphragms are molded with a stud in the center which is used for connection to the compressor or the valve stem. This imparts the flexing motion to the diaphragm as the valve stem is raised and lowered.

When plastic diaphragms such as TFE or Kel-F are used, a special compressor assembly is necessary because of the mechanical properties of these plastics. Table 4-5 shows the diaphragm materials available.

Some elastomeric diaphragm materials may be unique in their excellent resistance to certain chemicals at high temperatures. However, the mechanical properties of any elastomeric material will be lowered at the higher temperature with possible destruction of the diaphragm at high pressures. Consequently, the manufacturer should be consulted for his recommendation of allowable operating pressures when the operating temperature is above ambient.

All elastomeric materials operate best below 150°F, though some will function at higher temperatures. Viton, for example, is noted for its excellent chemical resistance and stability at high temperatures. However, when fabricated into a diaphragm, Viton is subject to lowered tensile strength just as any other elastomeric material would be at elevated temperatures. Fabric bonding strength also is lowered at elevated temperatures, and in the case of Viton, temperatures may be reached where the bond strength would become critical.

Fluid concentrations will also affect the diaphragm selection. Many of the diaphragm materials exhibit satisfactory corrosion resistance to certain corrodents up to a specific concentration and/or temperature. The elastomer may have a maximum temperature limitation based on mechanical properties which could be in excess of the allowable

---

[1] Registered Trademark of Victaulic Corp. of America.

Table 4-5. Diaphragm Materials

Valve Type	Material	Temperature Range, °F Min.	Max.
Weir	Soft natural rubber	−30	180
	Natural rubber	−30	180
	White natural rubber	0	160
	Pure gum rubber	−30	160
	Neoprene	−30	200
	Hi-temp black butyl	−20	250
	Hi-temp white butyl	−10	225
	Black tygon	0	150
	Clear tygon	0	150
	Hycar (general purpose) Buna-N	10	180
	Hypalon	0	225
	G.R.S.	−10	225
	TFE	−30	350
	Kel-F	60	250
	Polyethylene	10	135
	Ethylene propylene copolymer	−60	300
	Viton	−20	350
	TFE-faced	−60	300
Straightway	Natural rubber	−30	180
	Neoprene	0	180
	Hycar (Buna-N)	10	180
	Hi-temp black butyl	0	225
	Hi-temp white butyl	0	200
	Hypalon	0	200

operating temperature depending on its corrosion resistance. This should be checked from a corrosion table.

Stem and Bonnet Assembly

Diaphragm valves are available with indicating and nonindicating stem assemblies. Stems in diaphragm valves do not rotate. The handwheel rotates a stem bushing which engages the stem threads moving the stem up and down, raising and lowering the compressor which is pinned to the stem. The diaphragm in turn is secured to the compressor.

The indicating stem valve is identical to the nonindicating stem valve except that a longer stem is provided to extend up through the handwheel. See Fig. 4-28.

A quick opening bonnet, with lever operation, is also available. This bonnet is interchangeable with the standard bonnet on conventional

Courtesy of Grinnel Co.

Fig. 4-28. Diaphragm valve stem assemblies: A. Nonindicating stem; B. indicating stem; C. travel stop and indicating stem; D. lever operated.

weir-type bodies. A 90-degree turn of the lever moves the diaphragm from full open to full closed.

Many diaphragm valves are used in vacuum service. Standard bonnet construction can be employed in vacuum service on valves through 4 inches in size. On valves 6 inches and larger a sealed bonnet should be employed, and evacuated. This is recommended to guard against premature diaphragm failure.

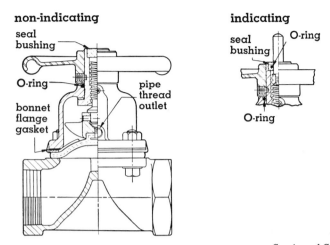

Fig. 4-29. Sealed bonnet construction.

Sealed bonnets are supplied with a seal bushing on the nonindicating types and a seal bushing plus O-ring on the indicating types. Construction is shown in Fig. 4-29.

This design of bonnet is also recommended for valves that are handling dangerous liquids and gases. In the event of a diaphragm failure the hazardous materials will not be released to the atmosphere. If the materials being handled are extremely hazardous, it is recommended that a means be provided to permit a safe disposal of the corrodents from the bonnet.

Diaphragm valves may also be equipped with chain wheel operators, extended stems, bevel gear operators, and air or hydraulic operators.

Identification of Parts

The nomenclature and identification of the components and parts of the diaphragm valves are shown in Fig. 4-30.

Installation, Operation, and Maintenance

General installation procedures and care of valves prior to installation are given starting on page 40. Covered here are the specific items unique to diaphragm valves.

When a diaphragm valve is used as an "unloading" valve (working pressure upstream and atmospheric or low pressure downstream), operation of the handwheel is relatively easy. On a "live line" application, where the back pressure on the valve is high, the valve becomes very

Ch. 4   GENERAL PURPOSE VALVES   109

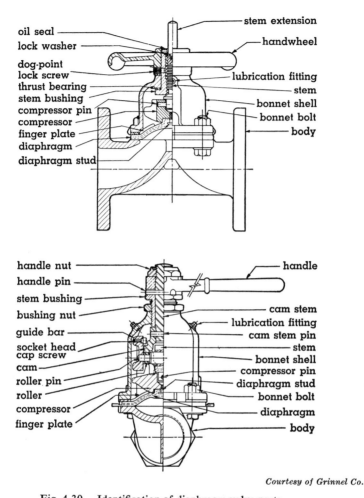

Courtesy of Grinnel Co.

Fig. 4-30. Identification of diaphragm valve parts.

difficult to operate, because of the thrust developed by the pressure acting over the entire diaphragm surface, which tends to reduce diaphragm life. This is illustrated by the example of a 3-inch valve installed in a 150-pound line. On a "live-line" application 1000 inch-pounds of torque are required for closure while the same valve on an "unloading line" requires only 480 inch-pounds of torque for closure. Similarly, a 4-inch valve requires 1600 inch-pounds of torque for closure on a live line while requiring only 850 inch-pounds for closure on an unloading line. Live-line pressure acting on the full diaphragm area in sizes of 10 inches and larger results in high stem loads. Consequently, the closing torque is higher than a single man can produce. For these larger valves where one man operation is required, the valve should be furnished with a manual gear operator. Under live-line closure, the diaphragm is being excessively worked, resulting in shortened diaphragm life.

One method of reducing closing torque is to utilize an air loaded bonnet assembly in the larger valve sizes (sealed bonnet). The bonnet is provided with seals in order to maintain air pressure without leakage. This bonnet can be appropriately connected to a pressure probe downstream of the valve, which in turn regulates air pressure to the bonnet of the valve so that the diaphragm is "balanced" to eliminate stress and to provide ease of closure by reducing the required closing force. This pressure loading scheme is particularly useful on valves equipped with diaphragm operators, such as in deionization applications where back pressures are usually high.

As can be appreciated, the diaphragm valve is, so to speak, a "positive" displacement valve, which displaces a volume of fluid equal to that volume swept by the diaphragm during closure. If the valve is in a portion of the line with valves shut off on both sides of the diaphragm valve, it is impossible to close the valve without damage. If this condition is not avoided through proper piping, and provisions made to prevent the condition, an operator may inadvertently apply a pipe wrench to the valve handwheel. As a result, the excessive internal pressure increases to a point where the diaphragm will rupture, or the compressor pin, stem, or sleeve may shear or bind. This possibility must be given particular thought where the valve utilizes solid plastic diaphragms in service, since the solid plastic diaphragms are limited in pressure rating and are more susceptible to failure from this type of abuse.

It is not necessary to apply bars, wrenches, "cheaters," or other aids in closing a diaphragm valve. A diaphragm valve is inherently a soft seating valve, and it is not necessary to grind hard, metal-to-metal seats together under peak torque in order to secure closure. In fact, the opposite is true, in that the resilient diaphragm is specifically designed for ease of seating to give satisfactory closure and to promote diaphragm life.

In many cases an operator will apply torques far in excess of those required to effect valve closure. This results in shortened diaphragm life and unnecessary expenses incurred in premature diaphragm replacements. If operating personnel are not familiar with diaphragm valves, or if they tend to be heavy handed in applying torques, the incorporation of travel stops is recommended during valve closure. These adjustable metal-to-metal travel stops, an optional feature available from the valve manufacturers, prevent the application of excessive torque on diaphragm closures.

Diaphragm valves are frequently used on services where a highly corrosive atmosphere is encountered. To resist atmospheric attack on the cast iron bonnets and handwheels, these parts are available with various protective coatings including Penton, epoxies, phenolics, or furanes. The small extra cost for these coatings will be quickly recovered on savings over frequent paint maintenance that normally is required and/or bonnet replacements if a continuing preventative maintenance program is not conducted.

In some applications the atmosphere may be particularly corrosive, which results in attack of the exterior and interior of the bonnet assembly, as well as on the valve lubricant. In such cases the bonnet assembly must be frequently maintained and lubricated. Alternately, it is possible to specify plastic-coated, or stainless steel, or monel or aluminum bonnet assemblies with stainless steel stems and sleeves. Such special bonnet assemblies can be lubricated with chemically inert grease such as the silicone lubricants or fluorocarbon lubricants.

Despite the fact that the diaphragm valve has an upstanding weir, it can be effectively drained for all intents and purposes if the valve is mounted so that the angle between the center line of the stem and the horizon is 15 to 18 degrees positive, depending upon the valve manufacturer. This brings the side tip of the weir down to a point where it nearly coincides with the floor of the pipe. At this point there is retained, at the most, 1 or 2 cubic centimeters of fluid. Alternately, where installation will permit, the valve can be installed vertically, where it will drain, except for fluid that has wetted the valve surfaces because of liquid surface tension.

Pressure grease fittings are supplied on diaphragm valve bonnets. It is recommended that a regular program of lubrication, four times a year, be carried out. It is important that valve bonnets be kept greased and in good working order to prevent the handwheel or sleeve assembly from "freezing." Such freezing may lead to the inadvertent application of large over-torque by the use of wrenches to operate the valve. This could result in broken bonnet flanges, broken handwheels, or seized stem and sleeve. Grease should be applied sparingly to keep the back of the diaphragm free of grease. Most elastomeric materials and polyethylene may be adversely affected by grease.

Only one part, the diaphragm, is normally subject to wear and needs replacement. Depending upon the type of service, it may last indefinitely.

If it is necessary to replace the diaphragm, the following steps should be followed:

Weir-Type Valves

1. Remove bonnet bolts and lift valve bonnet off body.
2. Remove diaphragm from compressor. (Follow manufacturer's instruction.)
3. Install new diaphragm. (Follow manufacturer's instructions.)
4. Replace bonnet on body and tighten bolts hand tight.
5. Close valve fully, back off one-quarter turn of handwheel, then tighten bolts with wrench.
6. Open valve and if necessary retighten bonnet bolts.

Straightway Valves

1. Remove bonnet bolts and lift bonnet off body.
2. Remove diaphragm from compressor. (Follow manufacturer's instructions.)
3. Install new diaphragm. (Follow manufacturer's instructions.)
4. With compressor holding diaphragm in slightly open position (one or two turns open from molded position), replace complete bonnet and tighten bolts securely.

Diaphragm Valve Specifications

A diaphragm valve specification should include the following items:

1. Body material
2. Diaphragm material
3. End connections
4. Type of stem assembly
   A. Nonindicating
   B. Indicating
5. Type of bonnet assembly
6. Type of operation
7. Operating pressure
8. Operating temperature.

4.8 PINCH VALVES

General

The relatively inexpensive pinch valve is the simplest in design of any valve. Actually it is an industrial version of the pinch cock used in the laboratory to control the flow of fluids through rubber tubing.

These valves are ideally suited for the handling of slurries, liquids with large amounts of suspended solids, and systems that convey solids pneumatically. Since the operating mechanism is completely isolated from the fluid, these valves also find application where corrosion or metal contamination of the fluid might be a problem.

Pinch valves are suitable for on-off and throttling services. However, the effective throttling range is usually between 10 percent and 95 percent of the rated flow capacity, a range that varies slightly among manufacturers.

Service Recommendations

1. Handling of slurry or fluids containing large amounts of suspended material
2. On-off and throttling services
3. Services with low maintenance cost
4. Low pressure drop through valve
5. Moderate temperature services.

Construction of Valve

The pinch valve consists of a sleeve molded of rubber or other synthetic material and a pinching mechanism. All of the operating portions are completely external to the valve. The molded sleeve is referred to as the valve body.

Since the natural and synthetic rubbers and plastics out of which the valve bodies are manufactured have good abrasion-resistance properties, and the flow through the valve is unimpeded, pinch valves are excellent for the handling of slurries, as mentioned previously. Many of these valves are used for handling mining slurries.

Sleeves are available with either expanded hubs and clamps designed to slip over a pipe end, or with a flanged end having standard ANSI dimensions.

The molded bodies, although reinforced with fabric, have moderate pressure and temperature limitations, depending upon the size of the valve. Average maximum operating temperature would be 250°F, with the average operating pressure varying depending upon the size, starting at 100 psig for 1-inch-diameter valves and decreasing to 15 psig for 12-inch-diameter valves. These figures are averages, and finite figures, depending upon the manufacturer, will vary considerably. Special valves have been designed and used with temperature ranges of minus 100° to plus 550°F. An operating pressure of 300 psig is not uncommon. These specifications should be verified by the manufacturer. Valve bodies are available in a wide range of materials. Some of the more common include:

Rubber	Neoprene	Buna-S
Hypalon	Hycar Buna-N	Silicone
Polyurethane	Viton-A	White neoprene
White rubber	Butyl	(food grade)
(food grade)		TFE

Most of the pinch valves are supplied with the sleeve (valve body) exposed. Another style fully encloses the sleeve within a metallic body. This type controls flow either with the conventional wheel and screw pinching device, or hydraulically or pneumatically with the pressure of the liquid or gas within the metal case forcing the sleeve walls together to shut off flow.

Most exposed sleeve valves have limited vacuum application because of the tendency of the sleeves to collapse when vacuum is applied. Some of the encased valves can be used on vacuum service by applying a vacuum within the metal casing and thus preventing the collapse of the sleeve.

Various types of pinch valves are shown in Fig. 4-31.

Courtesy of Flexible Valve Corp.

Fig. 4-31. Pinch valves.

Pinch Valve Specifications

Specifications for pinch valves should include the following items:

1. Operating pressure
2. Operating temperature
3. Material of construction of sleeve
4. Type of end connection
5. Exposed or encased sleeve.

Ch. 4 GENERAL PURPOSE VALVES 115

4.9 SLIDE VALVES

General

Slide valves are designed for handling liquids or gases containing a high percentage of solids, pulp stock, or free flowing granular materials. These valves do not supply a tight closure.

Flow is controlled by means of a gate or blade which slides between parallel body seats. There is no spreading mechanism. Closure is effected by fluid pressure forcing the downstream surface of the disc against the body seat.

Courtesy of Rockwell Mfg.

Fig. 4-32. Slide valve.

These valves are generally used for controlling low pressure flow of gases, liquids, suspensions, and fluidized solids — all applications where tight closure is not required. Since the blade is completely removed from the waterway when the valve is in the open position, there is very little pressure drop through the valve and very little obstruction to hamper the flow of solids.

Figure 4-32 shows a slide valve.

5

Check Valves

Check valves are designed to prevent the reversal of flow in a piping system. These valves are activated by the flowing material in the pipeline. The pressure of the fluid passing through the system opens the valve, while any reversal of flow will close the valve. Closure is accomplished by the weight of the check mechanism, by back-pressure, by a spring, or by a combination of the foregoing.

Basic styles of check valves have been designed for use with specific types of flow control valves, but do not necessarily have to be used with the corresponding flow control valve. When used with the corresponding flow control valve, the effect on flow of the check valve will be very similar to the effect on flow of the corresponding flow control valve.

The general types of check valves are:

1. Swing check valve
    A. Y-pattern
    B. Straight-through pattern
2. Tilting-disc check valve
3. Lift check valve
4. Piston check valve
5. Butterfly check valve
6. Spring-loaded check valve
7. Foot valve
8. Stop check valve.

Some check valves have been designed for specific applications, not necessarily related in any way to flow control valves. These conditions will be discussed under the heading of each specific valve. Table 5-1 lists the types of check valves and the corresponding style of flow control valve with which it is normally used.

CHECK VALVES

Table 5-1. Check Valve — Flow Control Valve Combinations

Type of Check Valve	Flow Control Valve Normally Used with:
Swing Check	Gate; Y-Pattern; Plug; Ball; Diaphragm
Tilting Disc	Gate; Y-Pattern; Plug; Ball; Diaphragm; Pinch
Lift Check	Globe, Angle
Piston Check	Globe, Angle
Butterfly Check	Butterfly; Plug; Ball; Diaphragm; Pinch
Spring Loaded *	Globe, Angle
Foot Valve	See note
Stop Check	See note

* These valves are designed for specific applications.

The considerations given to the selection of a check valve are very similar to the considerations given to the selection of a flow control valve. An analysis of the features of each check valve indicates a parallelism to the analysis of the features of a particular flow control valve which permits the formulation of the corresponding recommendations found in Table 5-1.

The size ranges and operating ranges of the different types of check valves are summarized in Table 5-2. The ranges shown are average and will vary among manufacturers. Maximum figures shown may not be attainable in all sizes and in all materials of construction.

General Installation and Maintenance

The general procedure for the care of valves prior to installation, found on page 40, applies to check valves as well as to flow control valves.

When installing check valves, make sure that all pipe strains are

Table 5-2. Sizes and Operating Ranges of Check Valves

	Size, Inches		Operating Ranges			
			Temp., °F		Pressure, Psig	
Check Valve	Min.	Max.	Min.	Max.	Min.	Max.
Swing Check						
Y-Pattern	¼	6	0	1250	0	2500
Straight-through	¼	24	0	1250	0	2500
Tilting Disc	2	30	−450	1100	0	1400
Lift Check	¼	10	0	1250	0	2500
Butterfly Check	1	72	0	500	0	1200
Spring Loaded	1	24	0	500	0	2500

kept off of the valves. The valves should not carry the weight of the line. Distortion from this cause results in inefficient operation and early maintenance. If the valve is of flanged construction it will be difficult to tighten the flanges properly. Piping should be supported by hangers placed on either side of the valve to take up the weight. Large heavy valves should be supported independently of the piping system so as not to induce a stress into the piping system.

Threaded-End Valves

Avoid undersize threads on the section of pipe on which the valve is to be installed. If the threaded section of pipe is too small, the pipe, when screwed into the valve to make a tight connection, may strike the seat and distort it so that the disc will not seat properly. Undersize threads on the pipe also make it impossible to get a tight joint. Safe practice is to cut threads to standard dimensions and standard tolerances. All pipe threads in valve bodies are gauged to standard tolerances.

Paint, grease, or joint-sealing compound should be applied only to the pipe (male) threads — *not* on the threads of the valve body. This reduces the chances of the paint, grease, or compound getting on the seat or other inner working parts of the valve to cause future trouble.

When installing screwed end valves always use the correct size wrenches with flat jaws (not pipe wrenches). By so doing there is less likelihood of the valve's being distorted or damaged. Also, the wrench should be used on the pipe side of the valve to minimize the chance of distorting the valve body. This is particularly important when the valve is constructed of a malleable material, such as bronze.

Flanged-End Valves

When installing flanged-end valves, tighten the flange bolts by pulling down the nuts diametrically opposite each other and in the numbered order as shown in Fig. 2-30 on page 42. All bolts should be pulled down gradually to a uniform tightness. Make all bolts finger tight first, then take three or four turns with a wrench on bolt number 1. Apply the same number of turns on each bolt, following the order shown in Fig. 2-30. Repeat the procedure as many times as required until the joint is tight. Uniform stress across the entire cross section of the flange eliminates a leaky gasket.

Socket-Weld End and Butt-Weld End Valves

The procedures for installing these check valves are the same as for installing the corresponding flow control valve and will be found on pages 42 and 43, respectively.

5.1 SWING CHECK VALVES

General

Swing check valves allow full unobstructed flow, opening with line pressure and automatically closing as pressure decreases, being fully closed when the line pressure reaches zero and thus preventing back flow. Turbulence and pressure drop within the valve are very low. Consequently, this style of check valve is normally recommended for use in systems employing gate valves.

Swing check valves can be installed in a horizontal line or vertical line with flow upward. In all cases the check valve must be installed with the flow pressure under the disc.

Service Recommendations

1. For minimum resistance to flow
2. For liquid service (low velocities)
3. For infrequent change of direction
4. For service in lines using gate valves.

Construction of Valve

Swing check valves are available in either a Y-pattern or straight-through body design. In either style the disc and hinge are suspended from the body by means of a hinge pin. Seating is either metal-to-metal or metal seat and composition disc. Composition discs are usually recommended for services where dirt or other particles may be present in the fluid, where noise is objectionable, or where positive shutoff at low pressure is required.

Y-pattern check valves are designed with an access opening in line with the seat, which is integral with the body. (Refer to Fig. 5-1.) This permits the disc to be rotated with a screwdriver to regrind the seating surfaces (on metal-to-metal seated valves) without removing the valve from the line.

Straight-through pattern check valves contain a disc that is hinged at the top (see Fig. 5-2). The disc seals against the seat which is integral with the body. This type of check valve usually has replaceable seat rings. The seating surface is placed at a slight angle to permit easier opening at lower pressures, more positive sealing, and less shock when closing under higher pressures.

When swing check valves are used in systems having frequent flow reversals, the valves may have a tendency to chatter. This can be corrected by equipping the valve with an outside lever and weight as shown in Fig. 5-3. The outside weight and lever serve a threefold purpose:

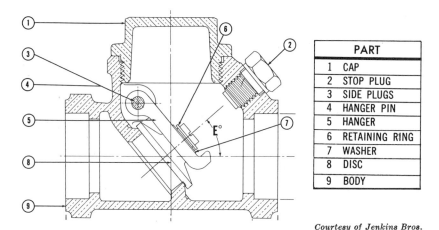

Fig. 5-1. Y-pattern swing check valve.

1. Weight is added to the disc for quicker closing against back flow, which prevents water hammer and excessive shock pressure.
2. Check valve cannot open until desired pressure is reached.
3. Operating sensitivity of the disc can be controlled.

When a check valve is operating properly there will normally be fluid trapped downstream of the valve. This is one purpose of installing a check valve. Such fluid can present a problem should maintenance have to be performed on the line, unless the fluid can be drained. To permit drainage of the line, most check valves can be supplied with a boss which may be drilled and tapped for a drain. If the piping system does not

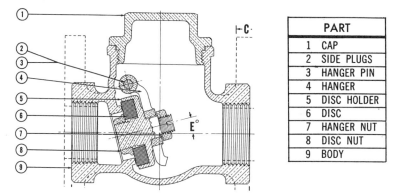

Fig. 5-2. Straight-through swing check valve.

For use in vertical lines for upward flow

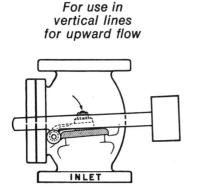

For use in horizontal lines

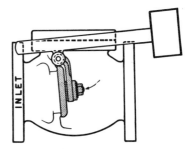

Courtesy of NIBCO Inc.

Fig. 5-3. Swing check valve with outside lever and weight. Right: Lever and weight are mounted to assist disc in closing.

contain an arrangement that will permit draining, then the check valve should be supplied with a drain connection. This is particularly important in piping systems handling corrosive and/or hazardous materials.

Identification of Parts

The nomenclature and identification of the components and parts of swing check valves are shown in Fig. 5-1 for the Y-pattern; in Fig. 5-2 for the straight-through pattern. Figure 5-3 illustrates the straight-through pattern with an outside lever and weight.

Available Materials of Construction

Swing check valves are available in a wide range of materials of construction. Included are the following body materials:

1. Bronze
2. All iron
3. Cast iron
4. Forged steel
5. Monel
6. Cast steel
7. Stainless steel.

Various trims are available, as are other metallic constructions. Some are standard and some can be had on special order, depending upon the specific manufacturer.

Installation and Maintenance

Swing check valves are usually installed in a system having gate valves because they provide comparable free flow. They are recommended for lines having low velocity flow and should not be used on lines with pulsating flow when the continual flapping or pounding would be destructive to the seating elements. This condition can be partially corrected by utilizing an external lever and weight as described previously.

Swing checks should be installed in horizontal lines or in vertical lines having upward flow. Pressure should always be under the seat. Be careful when installing a swing check valve. If you examine a swing check valve, you will notice that it has a plug on each side. These two plugs are your guide to correct installation. They hold in place the hanger pin which permits the disc to swing. The direction of swing is upward from the end of the valve *opposite* the plugs. Therefore, when installing a swing check valve, connect the end to which the side plugs are nearest to the *inlet flow* so that the incoming fluid will *open* the valve and the reverse flow will close the valve. If a swing check valve is installed in reverse position it will stop the flow. Some swing check valves have an arrow cast on the body to indicate the direction of flow. If this is the situation, pay attention to the arrow and the valve will be installed correctly.

If a swing check valve fails to seal against reverse flow, check the seating surfaces. Make sure that the line has been completely drained *before* removing the valve bonnet. If the seat is damaged or scored, it should be reground, or replaced if it is of the seat ring design. Inspect the disc. If the disc is of the composition type, verify that the disc material is compatible with the fluids being handled. Replace the composition disc with the correct materials. Before reassembling the valve, clean the internal portions thoroughly, being careful to remove all grinding compounds that may have been used if the seat was reground.

Other general instructions to be followed will be found on page 40.

5.2 TILTING DISC CHECK VALVE

The tilting-disc check valve is similar to the swing check valve. Like the swing check, the tilting-disc type keeps fluid resistance and turbulence low because of its straight-through design.

Tilting-disc check valves can be installed in horizontal lines and vertical lines having upward flow. Some designs simply fit between two flange faces providing a compact, lightweight installation, particularly in the larger diameters.

CHECK VALVES

Service Recommendations

1. For minimum resistance to flow
2. For liquid or gas service
3. For frequent change of direction
4. For service in lines using gate valves.

Construction of Valve

The disc lifts off of the seat to open the valve. The airfoil design of the disc allows it to "float" on the flow. Disc stops built into the body position the disc for optimum flow characteristics. A large body cavity helps minimize flow restriction. As flow decreases, the disc starts closing and seals before reverse flow takes effect. Back pressure against the disc moves it across the soft seal into the metal seat for tight shutoff, without slamming. If the reverse flow pressure is insufficient to effect a seal the valve may be fitted with an external lever and weight. Figure 5-4 illustrates the operation of the tilting-disc check valve.

These valves are available with either a soft seal ring (of an elastomeric or plastic material) and metal seat seal; or a metal-to-metal seal. The latter is recommended for high-temperature operation. The soft seal

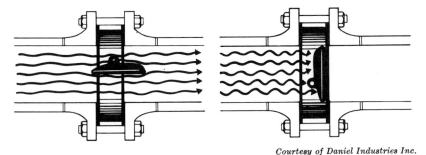

Courtesy of Daniel Industries Inc.

Valve Open

The contoured disc lifts easily off the seat to open the valve. The airfoil design of the disc allows it to "float" on the flow. Rugged disc stops built into the body position the disc for optimum flow characteristics. Large body cavity helps minimize flow restriction. The valve functions smoothly and silently in both horizontal and vertical up-flow lines.

Valve Closed

The disc has definite design advantages. It is counterweighted, mounted eccentrically and spring-loaded. As flow decreases, the disc starts closing and seals before reverse flow takes effect. Back pressure against the disc moves it across the soft seal into the metal seat for tight shut off.

Fig. 5-4. Operation of tilting-disc check valve.

rings are replaceable, but the valve must be removed from the line to make the replacement. The larger valves have an insert valve seat also.

Identification of Parts

The nomenclature and identification of the components and parts of the tilting-disc check valve are shown in Fig. 5-5.

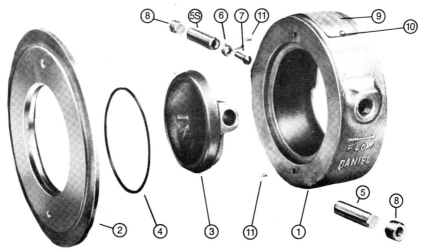

Courtesy of Daniel Industries Inc.

Fig. 5-5. Identification of parts of tilting-disc check valve: 1. Body, 2. seal housing, 3. disc, 4. soft seal, 5. hinge pin (without spring), 5S. hinge pin (slotted for spring), 6. spring, 7. spring retainer, 8. locking plug, 9. name plate, 10. name plate drive screw, 11. roll pin.

Available Materials of Construction

Tilting-disc check valves are available in the following materials of construction:

1. Carbon Steel
2. Cast Iron
3. Stainless Steel
4. Aluminum
5. Bronze.

Seal rings are available in Buna-N and TFE as standards, with other elastomeric materials available on order.

Installation and Maintenance

General installation instructions will be found on page 40.

Care should be taken when installing these valves that the inlet flow side is opposite the direction of the disc travel; otherwise the flow will be stopped. Arrows are usually cast on the body, and some valves have the word "inlet" cast on the inlet side of the disc.

These valves can be installed in horizontal lines or in vertical lines having upward flow.

If the valve is not sealing against reverse flow, drain the line, remove the valve and inspect the seal ring and the seat. If necessary, replace the seal ring, being sure that the material of construction of the seal ring is compatible with the materials being handled and the operating temperature.

5.3 LIFT CHECK VALVES

General

Lift check valves are commonly used in piping systems in which globe valves are being used as the flow control valve, since they have similar seating arrangements. Valves are available for installation in horizontal or vertical lines with upward flow.

They are recommended for use with steam, air, gas, water, and on vapor lines with high flow velocities.

Service Recommendations

1. For frequent change of directions
2. For steam, gas, or vapor service
3. Composition disc for air service
4. For use with globe and angle valves.

Construction of Valve

Lift check valves are available in three body patterns:

1. Horizontal
2. Angle
3. Vertical.

In design, the seat and disc configurations are very similar to the seat and disc configurations of globe valves, with the exception of the ball check valves which employ a ball as the disc. Flow to lift check valves must always enter below the seat. As the flow enters, the disc or ball is raised within guides from the seat by the pressure of the upward flow. When the flow stops or reverses, the disc or ball is forced onto the seat of the valve by both the backflow and gravity, effecting a seal. Some types of ball check valves may be installed horizontally. In this design

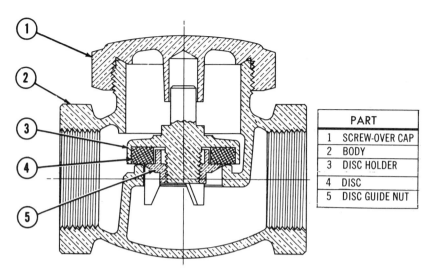

Courtesy of Jenkins Bros.

Fig. 5-6. Horizontal lift check valve.

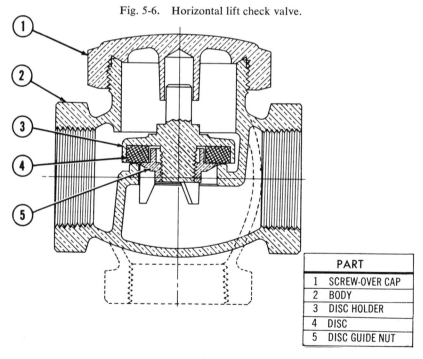

Courtesy of Jenkins Bros.

Fig. 5-7. Angle lift check valve.

Ch. 5 CHECK VALVES 127

the ball is suspended by a system of guide ribs. This type of check valve design is generally employed in all plastic check valves and/or impervious graphite check valves.

The seats of metallic body lift check valves are either integral with the body or contain renewable seat rings. Disc construction is similar to the disc construction of globe valves with either metal or composition discs. Metal disc and seat valves can be reground, the same as globe valves. Ball discs are generally recommended for the handling of viscous liquids.

Identification of Parts

The nomenclature and identification of parts and components of lift check valves are illustrated in Fig. 5-6 for horizontal lift check valves, Fig. 5-7 for angle lift check valves, and Fig. 5-8 for vertical lift check valves.

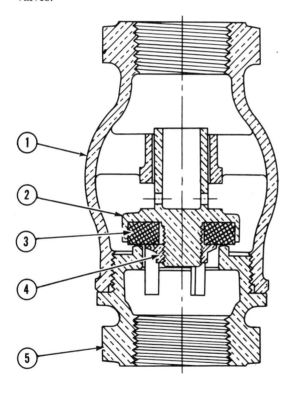

	PART
1	BODY
2	DISC HOLDER
3	DISC
4	DISC GUIDE NUT
5	SCREW-IN HUB

Courtesy of Jenkins Bros

Fig. 5-8. Vertical lift check valve.

Available Materials of Construction

Lift check valves are not only available in a wide range of metallic constructions but also in various plastics and other materials. A representative list of standard materials of construction in which lift check valves are available includes:

Bronze
All iron
Cast iron
Forged steel
Monel
Stainless steel

Polyvinyl chloride (PVC)
Chlorinated polyether (penton)
Impervious graphite
Borosilicate glass — TFE
TFE-lined

with a variety of trim combinations.

Figure 5-9 shows the lift check valves in borosilicate glass — TFE combination and TFE-lined construction.

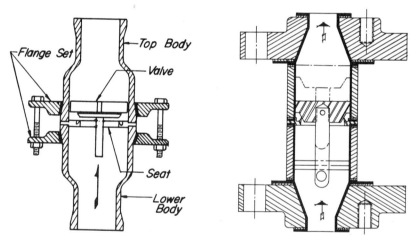

Courtesy of Chem Flow Corp.

Fig. 5-9. Borosilicate glass — TFE (left) and TFE-lined (right) lift check valves.

Installation and Maintenance

General installation procedures to follow for check valves will be found on page 40. When installing lift check valves, certain other factors must be taken into consideration.

Horizontal life checks should be installed in horizontal lines with operating parts in the vertical position and pressure under the seat, assuring the return of the disc to its seat for positive closing.

The angle pattern must be installed with flow upward from beneath the seat. The vertical pattern is for use on vertical pipes only and must also have the flow upward, from beneath the seat.

If the valves do not prevent backflow, drain the line, dismantle the valve, and inspect the seat and disc. Replace damaged or scored parts and reinstall. On valves with regrindable seats and discs, regrind. Make sure that all grinding compound is removed from the valve before reinstalling. Clean the interior of the valve body whether the seat has been reground or not. If composition discs are being replaced, make sure that the composition disc is compatible with the media in the pipeline.

5.4 PISTON CHECK VALVES

General

Piston check valves are essentially lift check valves, with a dashpot consisting of a piston and cylinder that provides a cushioning effect during operation. Because of the similarity in design to lift check valves, the flow characteristics through a piston check valve are essentially the same as through a lift check valve.

Installation is the same as for a lift check in that the flow must enter from under the seat. Seat and disc construction is the same as for lift checks.

Piston checks are used primarily in conjunction with globe and angle valves in piping systems experiencing very frequent changes in flow direction.

5.5 BUTTERFLY CHECK VALVES

Butterfly check valves have a seating arrangement similar to the seating arrangement of butterfly valves. Flow characteristics through these check valves are likewise similar to the flow characteristics through butterfly valves. Consequently, butterfly check valves are quite frequently used in conjunction with butterfly valves. In addition the construction of the butterfly check valve body is such that ample space is provided for unobstructed movement of the butterfly valve disc within the check valve body without the necessity of installing spacers. (See page 100.)

As with butterfly valves the basic body design lends itself to the installation of seat liners of many materials of construction. This permits the construction of a corrosion-resistant valve at less expense than would be encountered were it necessary to construct the entire body of the higher alloy or more expensive metal. This is particularly true in constructions such as those of titanium.

Since the flow characteristics are similar to the flow characteristics of butterfly valves, applications of these valves are much the same. Also, because of their relatively quiet operation they find application in the air conditioning and heating systems of high rise buildings. Simplicity of design also permits their construction in large diameters — up to 72 inches.

Service Recommendations

1. For minimum resistance to flow
2. For frequent change of direction
3. For liquid or gas service
4. For use in lines using butterfly valves.

Construction of Valve

The butterfly check valve design is based on a flexible sealing member sealing against the bore of the valve body at an angle of 45 degrees. The short distance the discs must move from full open to full closed inhibits the "slamming" action found in some other types of check valves. Figure 5-10 shows the internal assembly of the butterfly check valve.

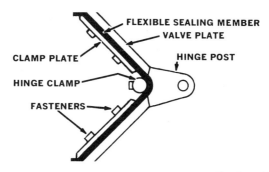

Courtesy of Techno Corp.

Fig. 5-10. Internal assembly of butterfly check valve.

Flexible sealing members are available in Buna-N, Neoprene, Nordel, Hypalon, Viton, Tygon, Urethane, Butyl, Silicone, and TFE as standard, with other materials available on special order.

The valve body essentially is a length of pipe which is fitted with flanges or has threaded, grooved, or plain ends. The interior is bored to a fine finish. The flanged end units can have liners of various metals or plastics installed depending upon the service requirement. Internals and fasteners are always of the same material as the liner. Figure 5-11 illustrates such a lined valve.

CHECK VALVES

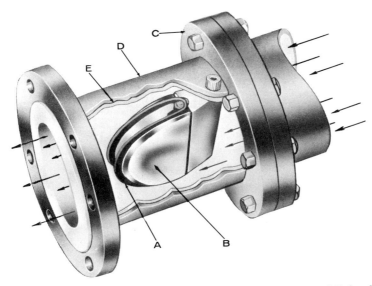

Courtesy of Techno Corp.

Fig. 5-11. Lined butterfly check valve. A. Sealing member as required, B. internals to match liner, C. ASA flanges, D. steel or iron body, E. liner.

A short form of this check valve is also available which fits between pipe flanges in the same manner as the wafer butterfly valve. See Fig. 5-12.

Available Material of Construction

These valves are available in a fairly wide range of materials of construction, particularly in the lined styles. Standard body fluid contact parts (either in lined or solid construction) include:

Steel	Polyvinyl chloride (PVC)	Cast iron
Stainless steel	Polyvinyldichloride (CPVC)	Monel
Titanium	Polyethylene	Bronze
Aluminum	Polypropylene	

Flexible sealing members are available in the following materials as standard, with others available on special order:

Buna-N	Neoprene	Nordel
Viton	Hypalon	Tygon
Butyl	Urethane	Silicone
TFE		

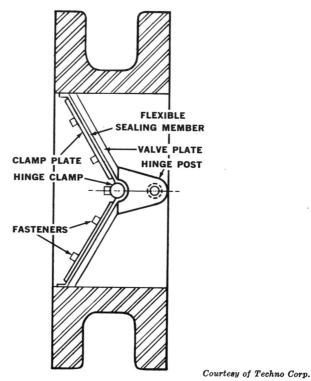

Fig. 5-12. Wafer-type butterfly check valve.

Installation and Maintenance

Butterfly check valves may be installed in any position — horizontal or vertical with the vertical flow either upward or downward. On lined valves care should be taken to protect the valve liner from damage during handling. Severe scratches or damage to the liner on the flange face may not be overcome by flange pressure, making a gasket seal impossible.

Care should be taken to insure that the valve is installed so that the entering flow comes from the hinge post end of the valve; otherwise all flow will be stopped. Additional general installation procedures will be found on page 40.

5.6 SPRING-LOADED CHECK VALVES

General

Spring-loaded check valves are actually a form of vertical lift check valves — the difference being that a spring is employed to effect the sealing. Advantages of spring actuation are: more rapid response to reversal of flow (usually closes as soon as the fluid velocity has reached zero and before reverse flow has actually commenced) and the ability to be installed in any position vertical or horizontal. Because of their rapid closing action they are more effective in controlling hydraulic shock in a piping system than some other types of check valves.

This type of construction also permits the furnishing of check valves in materials of construction that would not be possible if it were necessary to rely upon gravity and liquid head to effect the seal.

Service Recommendations

1. For frequent reversal of flow
2. For liquid or gas service
3. For positive sealing
4. For controlling hydraulic shock in a pipeline.

Construction of Valve

Details of construction vary somewhat among valve manufacturers, but in general the seat and disc configuration resembles that of a globe valve with a spring arrangement to hold the disc on the seat until sufficient pressure is developed to open the valve. Seats and discs are replaceable.

Available Materials of Construction

Spring-loaded check valves are available with body materials of practically any material of construction, including bronze, steel, stainless steel, borosilicate glass, TFE-lined, various plastic materials, etc. There is a wide choice of trim materials, including bronze, stainless steel, TFE, various plastics, etc.

Figure 5-13 illustrates a steel valve with stainless steel trim; and Fig.

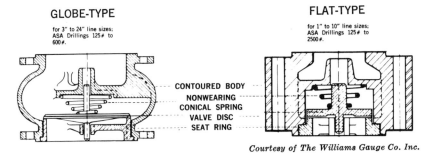

Fig. 5-13. Steel-body spring-loaded check valve.

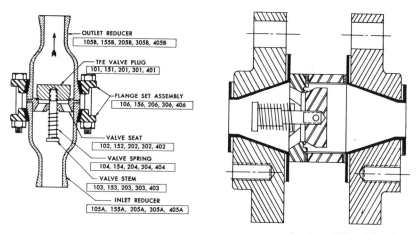

Courtesy of Chem Flow Corp.

Fig. 5-14. Borosilicate glass and TFE-lined spring-loaded check valve.

5-14 shows spring-loaded check valves with a borosilicate glass body and a TFE-lined body, both with TFE trim.

Installation and Maintenance

The general principles of check valve installation (see page 117) should be followed. The spring-loaded check valves can be installed in horizontal, angular or vertical lines. The flow must enter under the seat but these valves can be installed in vertical downflow as well as vertical upflow positions.

Seats and discs are replaceable; therefore in the event of leakage the seats and discs should be inspected for damage and replaced if necessary.

5.7 STOP CHECK VALVES

The stop check valve is a check valve which can manually or through a motor-operated stem tightly seal the check valve or limit its opening. The valve stem is not connected to the disc. When the stem is in the open position, such a valve operates as a conventional lift check valve. However, the stem feature permits shutting off, or throttling of, the normal flow through the valve. The closure against reverse flow feature is always present. Figure 5-15 illustrates such a valve.

5.8 FOOT VALVES

Foot valves are basically lift check valves which are used on the bottom of a suction line to maintain the pump's prime. They may be of the lift

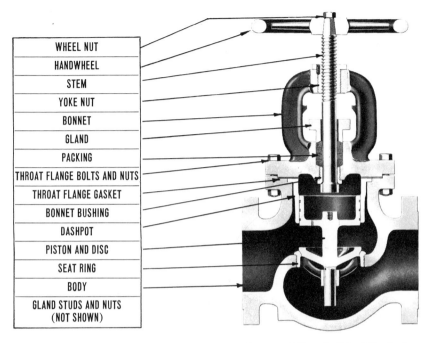

WHEEL NUT
HANDWHEEL
STEM
YOKE NUT
BONNET
GLAND
PACKING
THROAT FLANGE BOLTS AND NUTS
THROAT FLANGE GASKET
BONNET BUSHING
DASHPOT
PISTON AND DISC
SEAT RING
BODY
GLAND STUDS AND NUTS (NOT SHOWN)

Courtesy of Kennedy Valve Mfg. Co. Inc.

Fig. 5-15. Stop check valve.

check or ball check variety and are usually fitted with a strainer. The head of liquid above the valve holds the valve closed and keeps the suction pipe full. As suction is created when the pump starts up, the check valve opens and permits flow to the pump.

6

Special Service Valves

This section is concerned with the more common valves which have been designed for specific applications. Some of the valves are operated manually in much the same manner as stop valves while others are automatic in operation. Since it is not the intent of this book to deal with automatic control valves which are components in a control circuit, the only automatically operating valves discussed are those which are self-contained, such as pressure reducing valves. The various valves included in this section are:

Pressure relief valves   Pressure reducing valves
Flush bottom valves   Backpressure-regulating valves
Sampling valves   Cryogenic valves
Solenoid valves

Other valves, such as jacketted valves, are also available, but these are variations external to the valve itself. Most styles of valves are available with jackets to permit heating of the body so that solidification will not take place where materials having high melting points are handled at increased temperatures.

6.1 PRESSURE RELIEF VALVES

General

The terms "safety valve" and "relief valve" are quite often used interchangeably to designate valves that protect against excessive pressure. However, there is a difference between them.

Safety valves are designed to have a full opening pop action to provide immediate relief. These valves are required by Section VIII of the ASME code. Capacities, overpressure, and blowdown for safety valves are specified in the code.

Relief valves are designed to open slowly with increases in the initial pressure. These valves do not have a code design. Relief valves are normally used to relieve excessive pressures developed by noncompressible fluids since a relatively small discharge of a noncompressible fluid will provide immediate relief. Under these conditions it is not necessary that the relief valve open immediately to a full open position, but rather that it continue to open as pressure increases.

Because of these differences in action, safety valves are usually employed to relieve excessive pressure build-up caused by gases (compressible fluids), while relief valves are used to relieve excessive pressure build-up caused by noncompressible fluids.

Safety-relief valves have a full opening pop design and may be used on either compressible or noncompressible fluids.

Definitions

There are certain terms which must be understood when sizing, selecting, and specifying pressure relief valves. The following terms are the ones most frequently encountered:

Pressure relief valve. A generic term applying to relief valves, safety valves, or safety relief valves, designating a valve which is an automatic pressure-relieving device.

Safety valve. An automatic pressure-relieving device actuated by the static pressure upstream of the valve, and characterized by rapid full opening or pop action. It is used for steam, gas, or vapor service.

Relief valve. An automatic pressure-relieving device actuated by the static pressure upstream of the valve, which opens in proportion to the increase in pressure over the opening pressure. It is used primarily for liquid service (noncompressible fluid).

Safety relief valve. An automatic pressure relieving device activated by the static pressure upstream of the valve, suitable for use as either a safety or relief valve depending upon application.

Set pressure.

    1. *Liquid service:* In a relief valve or safety relief valve on liquid service, the set pressure in pounds per square inch gauge is to be considered the inlet pressure at which the valve starts to discharge under service conditions.

    2. *Gas, vapor, or steam service:* In a safety or safety relief valve on gas, vapor, or steam service, the set pressure, in pounds per

square inch gauge, is to be considered the inlet pressure at which the valve pops open under service conditions.

Differential set pressure. The pressure differential in pounds per square inch between the set pressure and the constant superimposed back pressure. It is applicable only when a conventional-type safety relief valve is being used in service against a constant superimposed back pressure.

Cold differential test pressure. The pressure, in pounds per square inch gauge, at which the valve is adjusted to open on the test stand. This cold differential test pressure includes the corrections for service conditions of back pressure and/or temperature.

Operating pressure. The operating pressure of the vessel is the pressure, in pounds per square inch gauge, to which the vessel is usually subjected in service. A vessel is usually designed for a maximum allowable working pressure, in pounds per square inch gauge, which will provide a suitable margin above the operating pressure in order to prevent undesirable operation of the relief device.

Maximum allowable working pressure. All vessels operating in excess of 15 psig, and/or jackets on vessels operating in excess of 15 psig, must be designed and constructed in accordance with the ASME code, Section VIII, for Unfired Pressure Vessels. Such vessels have a name plate indicating the maximum allowable working pressure of the vessel coincident with a maximum allowable operating temperature. The vessel may not be operated above these set conditions; and consequently, this is the highest pressure at which the primary pressure relief valve is set to open.

Since operating temperature affects the allowable operating pressure, a reduction in temperature would permit an increase in operating pressure. However, Section VIII of the ASME code would have to be checked to determine the allowable increase. Conversely, an increase in operating temperature will result with a reduction in the allowable operating pressure.

Overpressure. Pressure increase, over the set pressure of the primary relieving device is overpressure. It is the same as accumulation only when the relieving device is set at the maximum allowable working pressure of the vessel.

Accumulation. Pressure increase over the maximum allowable working pressure of the vessel during discharge through the pressure relief valve, expressed as a percentage of that pressure, or pounds per square inch, is called accumulation.

Blowdown. Blowdown is the difference between the set pressure and the resetting pressure of a pressure relief valve, expressed as a percentage of the set pressure, or pounds per square inch.

Ch. 6 SPECIAL SERVICE VALVES 139

Lift. The rise of the valve disc in a pressure relief valve is called lift.

Backpressure. Pressure on the discharge side of a pressure relief valve, which can be either constant or variable.

    1. *Constant backpressure:* Backpressure which does not change appreciably under any condition of operation whether the pressure relief valve is open or closed.

    2. *Variable backpressure:* Backpressure which develops as a result of the following conditions:

        A. *Built-up backpressure:* The pressure in the discharge header which develops as a result of flow after the pressure relief valve opens.

        B. *Superimposed backpressure:* The pressure in the discharge header before the relief valve opens.

Sizing

The first step is to calculate the required flow rate which the pressure relief valve must pass in order to prevent excessive accumulation. In exothermic types of reactions the valve must be sized to pass a flow rate capable of providing pressure relief at the maximum rate at which the pressure can develop.

After the fluid capacity to be relieved has been determined, it is necessary to calculate the orifice area required to relieve the predetermined quantity of liquid or vapor. Once the area has been established, the actual valve selection can be made by consulting manufacturers' tables which will list several valves of the required orifice area. Final selection will be based on the matching of the orifice area with the valve or valves which meet the pressure, temperature, and materials of construction requirements. Solution of the following formulas will give the effective orifice area that will be required to handle the specified fluid capacity:

$$\frac{For\ Vapors\ or\ Gases}{(lbs/hr)} - \text{ASME-UPV Formula}$$

$$A = \frac{W\sqrt{T}\sqrt{Z}}{CKP\sqrt{M'}\,K_b}$$

$K_b = 1$ when backpressure is below 55 percent of absolute relieving pressure.

$$\frac{For\ Vapors\ or\ Gases}{(S.C.F.M.)} - \text{Converted ASME-UPV Formula}$$

$$A = \frac{V\sqrt{G}\sqrt{T}\sqrt{Z}}{1.175 CKPK_b}$$

$K_b = 1$ when backpressure is below 55 percent of absolute relieving pressure.

$\dfrac{\textit{For Steam}}{\text{(lbs/hr)}}$ — ASME-UPV Formula

$$A = \dfrac{W_s}{51.5 K P K_b K_{sh}}$$

$K_b = 1$ when backpressure is below 55 percent of absolute relieving pressure.

$K_{sh} = 1$ for saturated steam.

$\dfrac{\textit{For Liquids}}{\text{(GPM)}}$

$$A = \dfrac{V_L \sqrt{G}}{27.2 P_d K_p K_u}$$

$K_p = 1$ at 25 percent overpressure
$K_u = 1$ at normal viscosities

$\dfrac{\textit{For Air}}{\text{(S.C.F.M.)}}$

$$A = \dfrac{V_a \sqrt{T}}{418 K P K_b}$$

$K_b = 1$ when backpressure is below 55 percent of absolute relieving pressure.

In the foregoing formulas the following nomenclature applies:

A = Required orifice area in square inches.
W = Required vapor capacity in pounds per hour.
W_s = Required steam capacity in pounds per hour.
V = Required gas capacity in S.C.F.M. (standard cubic feet per minute).
V_a = Required air capacity in S.C.F.M.
V_L = Required liquid capacity in U.S. gallons per minute.
G = Specific gravity of gas (air = 1) or liquid (water = 1) at actual discharge temperature. (A specific gravity at any lower temperature will obtain a safe valve size.)
M = Average molecular weight of vapor.
P_d = Differential set pressure in pounds per square inch (set pressure minus constant or maximum variable backpressure, psig).
P = Relieving pressure in psia. = set pressure + overpressure + 14.7.

T = Inlet temperature °R (°F plus 460).

Z = Compressibility factor corresponding to T and P. (If this factor is unknown, compressibility correction can be safely ignored by using a value of $Z = 1.0$.)

C = Gas or vapor flow constant. Select from Table 6-1 using ratio of specific heats (k) of vapor or gas being considered.

k = Ratio of specific heats, C_p/C_v. This value is a constant for an ideal gas. If this ratio is not known, the value $k = 1.001$; $c = 315$ will result in a safe valve size. Isentropic coefficient (n) may be used instead of (k).

K_p = Vapor or gas flow correction factor for constant backpressure above critical pressure.

K_W = Liquid flow factor for variable backpressure.

K_v = Vapor or gas flow factor for variable backpressure.

K_{sh} = Super heat correction factor.

K_u = Viscosity correction factor.

K = Coefficient of discharge.

The coefficient of discharge and the various correction factors should be supplied by the valve manufacturer for his specific valves.

Reactive Force

The discharge of a pressure relief valve with unsupported piping will impose a reactive load on the inlet of the valve because of the reaction force of the flowing fluid. This is particularly important where

Table 6-1. Constant C for Gas or Vapor Related to Specific Heats
($k = C_p/C_v$)

k	C	k	C	k	C
1.00	315	1.26	343	1.52	366
1.02	318	1.28	345	1.54	368
1.04	320	1.30	347	1.56	369
1.06	322	1.32	349	1.58	371
1.08	324	1.34	351	1.60	372
1.10	327	1.36	352	1.62	374
1.12	329	1.38	354	1.64	376
1.14	331	1.40	356	1.66	377
1.16	333	1.42	358	1.68	379
1.18	335	1.44	359	1.70	380
1.20	337	1.46	361	2.00	400
1.22	339	1.48	363	2.20	412
1.24	341	1.50	364		

piping discharging to atmosphere includes a 90-degree turn and has no support for the outlet piping. All reactive loading due to the operation of the valve is then transmitted to the valve and inlet piping. The following formula is based on the assumptions that critical flow of the gas or vapor is obtained at the outlet of the safety-relief valve and that the discharge is horizontal to the atmosphere, without discharge piping. Under conditions of subcritical flow at the outlet, the actual force may be less than the force computed from this formula. Subcritical flow may occur at the outlet at low set pressures or where the safety-relief valve discharges into a closed system. The formula is derived from the law of momentum.

For any gas or vapor:

$$F = \frac{W\sqrt{\frac{kT}{(k+1)M}}}{366}$$

where:

F = Horizontal reactive force at center line of valve when any gas or vapor is flowing, pounds
W = Flow of any gas or vapor, pounds per hour
k = Ratio of specific heats C_p/C_v
M = Molecular weight of any gas or vapor
T = Absolute temperature at inlet (degrees F + 460).

Provision should be made to resist this force.

Valve Construction

Safety and relief valves are normally held in the closed position by means of a spring loaded disc. The spring loading is adjusted so that a predetermined pressure acting under the valve disc (seat) will lift the disc from the seat permitting fluid to pass through.

In safety valves the disc overhangs the seat, to provide additional thrust area after the initial opening, and therefore faster rise of the disc to the full open position. The seat is usually surrounded by an adjustable ring (or less frequently, the disc is built with a skirt) to form a huddling chamber so that once the valve begins to open, the pressure is also applied to the additional exposed surface and not to the disc alone. Refer to Fig. 6-1.

By adjusting this ring the blowdown pressure, which is the difference between the relief pressure and the slightly lower pressure of the system at which the valve closes, may be regulated. Care must be taken in making these adjustments not to decrease the blowdown pressure exces-

Ch. 6 SPECIAL SERVICE VALVES 143

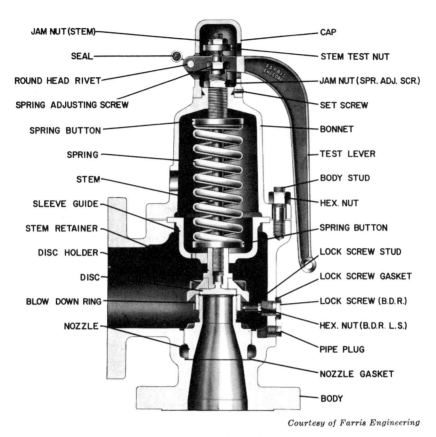

Courtesy of Farris Engineering

Fig. 6-1. Safety valve.

sively since a small blowdown may cause the valve to chatter and prevent it from opening quickly.

Relief valves are designed so that the area exposed to the overpressure is the same whether the valve is open or closed. Because of this the valve disc is lifted from its seat gradually as pressure increases until the valve reaches the full open position. Refer to Fig. 6-2.

The majority of safety valves are spring loaded. However, there are some safety valves which have an external lever and weight. Varying set pressures are established by the distances of the weight on the lever from the valve stem.

Spring-loaded pressure-relief valves can have their set pressure altered by means of a screw in the top of the bonnet which adjusts the spring compression. See Figs. 6-1 and 6-2.

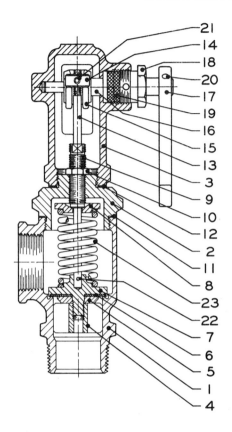

ITEM	PART NAME
1	BODY
2	BONNET
3	CAP
4	DISC GUIDE
5	DISC RETAINER
6	DISC SEAT
7	DISC HOLDER
8	SPRING BUTTON-UPPER
9	SPRING ADJUSTING SCREW
10	JAM NUT
11	BONNET GASKET
12	CAP GASKET
13	STEM
14	STEM TEST NUT
15	CAM
16	CAM SHAFT
17	TEST LEVER
18	GLAND
19	PACKING RING
20	GROOV-PIN
21	SET SCREW
22	SPRING PIN
23	SPRING

Courtesy of Farris Engineering

Fig. 6-2. Relief valve.

The spring on pressure relief valves is located in the bonnet. In some designs the entire bonnet area is open to contact to the media being passed through the valve when the valve opens. For applications where the material being handled is noncorrosive, and/or will not have any effect on the performance of the valve, this construction is satisfactory. However, if the media being handled is corrosive (to the spring construction), sticky, or dirty, then a diaphragm construction should be employed. In these valves a diaphragm of a suitable resistant material seals the bonnet from contact with the process media. The diaphragm flexes as the disc operates.

Pressure relief valves are available from cryogenic temperatures up to approximately 1500°F and from full vacuum up to 10,000 psi.

Most safety valves, and some relief valves, are equipped with an external relief-checking lever.

Available Materials of Construction

Pressure relief valves are available in a range of materials including:

Cast iron	Bronze	Stainless steel
Carbon steel	Brass	Hastelloy
Glass-TFE	TFE-lined	Monel

Trim comes in varying materials. (Trim on a pressure relief valve generally refers to the internal parts.)

Installation and Maintenance

The method of installation of pressure relief valves should be in accord with the provisions of the ASME Unfired Pressure Vessel Code: Section VIII, paragraphs UG-125, UG-126, UG-128, UG-134, and appendix M. Paragraphs UG-132 and UG-133 and appendix J should be consulted when pressure relief valves are to be sized. The code should be studied and understood when pressure relief valves are being sized, selected, and installed. Following are excerpts from the code which deal specifically with installation of pressure relief valves.

1. Pressure relief valves should be located and installed so that they are readily accessible for inspection and repair.

2. If the design of a safety or relief valve is such that liquid can collect on the discharge side of the disc, the valve shall be equipped with a drain at the lowest point where liquid can collect.

3. The spring in a safety relief valve in service for pressures up to and including 20 psi shall not be reset for any pressure more than 10 percent above or 10 percent below that for which the valve is marked. For higher pressures, the spring shall not be reset for any pressure more than 5 percent above or 5 percent below that for which the safety or relief valve is marked.

4. No liquid relief valve is to be less than ½-inch iron pipe size.

5. Safety and relief valves shall be connected to the vessel in the vapor space above any contained liquid, or to piping connected to the vapor space in the vessel which is to be protected.

6. The opening through all pipe and fittings between a pressure vessel and its pressure-relieving device shall have at least the area of the pressure-relieving device inlet, and in all cases shall have sufficient area so as not to unduly restrict the flow to the pressure relieving device.

7. When two or more required pressure-relieving devices are placed on one connection, the internal cross-sectional area of this inlet connection shall be at least equal to the combined inlet areas of the safety devices

connected to it, and in all cases shall be sufficient so as not to restrict the combined flow of the attached devices.

8. Liquid relief valves should be connected below the normal liquid level.

9. There shall be no intervening stop valves between the vessel and the pressure relief valve, or between the pressure relief valve and the point of discharge except:

A. when these stop valves are so constructed or positively controlled that the closing of the maximum number of block valves possible at one time will not reduce the pressure-relieving capacity provided by the unaffected relieving devices below the required relieving capacity, or

B. under the considerations set forth in 10 and 11.

10. A vessel, in which pressure can be generated because of service conditions, may have a full-area stop valve between it and its pressure-relieving device for inspection and repair purposes only. When such a stop valve is provided, it shall be so arranged that it can be locked or sealed open, and it shall not be closed except by an authorized person who shall remain stationed there during that period of the vessel's operation within which the valve remains closed, and who shall again lock or seal the stop valve in the open position before leaving the station.

11. A full area stop valve may be placed on the discharge side of a pressure relieving device when its discharge is connected to a common header with other discharge lines from other pressure-relieving devices on nearby vessels that are in operation, so that this stop valve, when closed, will prevent a discharge from any connected operating vessels from backing up beyond the valve. Such a stop valve shall be so arranged that it can be locked or sealed in either the open or closed position, and it shall be locked or sealed in either position only by an authorized person. Under no condition should this valve be closed while the vessel is in operation except when a stop valve on the inlet side of the safety-relieving device is installed and is first closed as in item 10.

12. Discharge lines from pressure relief valves should be designed to facilitate drainage, or should be fitted with an open drain to prevent liquid from lodging in the discharge side of the valve. All such lines should lead to a safe place of discharge. The discharge line should be sized such that any pressure which may exist or develop will not reduce the relieving capacity of the valve below that required to protect the vessel.

13. Where practical, the use of a short discharge pipe, or vertical riser, connected through long-radius elbows from each individual pressure-relief valve, blowing directly to atmosphere is recommended. Such pipes

should be at least of the same size as the valve outlet. Where the nature of the discharge permits, broken discharge lines are recommended since condensed vapor in the discharge line, or rain, may be collected in a drip pan and piped to a drain. This construction has the additional advantage of not transmitting discharge pipe strains to the valve. In this type of installation, the backpressure effect will be negligible and no undue influence upon normal valve operation can result.

14. When discharge lines are long, or where outlets of two or more valves having set pressures within a comparable range are connected into a common line, the effect of the backpressure that may be developed when certain valves operate must be considered. The sizing of any section of a common discharge header downstream from each of two or more pressure relief valves that may reasonably be expected to discharge simultaneously shall be based on the total of their outlet areas, with due allowance for the pressure drop in all downstream sections. For such service, pressure relief valves, specifically designed for use on high or variable backpressure, should be used. Manufacturers of this type of valve should be consulted for sizing and recommendations.

15. All discharge lines should be run as direct as possible to the point of final release. For longer lines, long-radius elbows should be used when changes in direction are necessary. Care should be taken in the design of the line to avoid close-up fittings and to minimize line strains (use expansion joints), line-sway, and vibration.

16. It is essential that a routine preventative maintenance inspection program be followed for each pressure relief valve. Each and every pressure relief valve should be inspected *and tested* periodically to insure its safe operation. The intervals between inspection and test will vary for each valve depending upon its service. Valves on noncorrosive, clean service should be inspected *and tested* at least once a year. Valves on corrosive service should be inspected more frequently, the time interval being dependent upon the severity of the corrodent.

Records should be maintained of these inspections and tests so that a check can always be made as to when the last time was that the valve had been inspected.

Capacity Conversions for Safety Valves

Whenever a processing vessel is to be used in a service other than what it had originally been designed for, the pressure relief valve should be investigated. A common mistake is to check the pressure relief valve set pressure only. This is not sufficient. A check must also be made as to its relieving capacity under the new conditions.

The capacity of a safety or relief valve in terms of a gas or vapor other than the medium for which the valve was officially rated may be determined by application of the formulas given.

Knowing the official capacity of a safety valve, which is stamped on the valve, it is possible to determine the overall value of KA in either of the following formulas when the values of these individual terms are not known:

Official rating in steam

$$KA = \frac{W_s}{51.5P}$$

Official rating in air

$$KA = \frac{W_a}{CP}\sqrt{\frac{T}{M}}$$

The value of KA is then substituted in the following formulas to determine the capacity of the safety valve in terms of the new gas or vapor:

For steam

$$W_s = 51.5KAP$$

For air

$$W_a = CKAP\sqrt{\frac{M}{T}}$$

$$C = 356$$

$$M = 28.97$$

$T = 520$ when W_a is the rated capacity.

For any gas or vapor

$$W = CKAP\sqrt{\frac{M}{T}}$$

where:
- W_s = Rated capacity, pounds of steam per hour
- W_a = Rated capacity, converted to pounds of air per hour at 60°F inlet temperature
- W = Flow of any gas or vapor, pounds per hour
- C = Constant of any gas or vapor which is a function of the ratio of specific heats, $k = C_p/C_v$ (See Table 6-1.)
- K = Coefficient of discharge
- A = Actual discharge area of the safety valve, square inches

P = (Set pressure × 1.10) plus atmospheric pressure, pounds per square inch absolute
M = Molecular weight
T = Absolute temperature at inlet (°F + 460).

These formulas may also be used when the required flow of any gas or vapor is known and it is necessary to compute the rated capacity of steam or air.

For hydrocarbon vapors where the actual value of k, the ratio of specific heats (C_p/C_v), is not known the conservative value $k = 1.001$ has been commonly used and the formula becomes:

$$W = 315KAP\sqrt{\frac{M}{T}}$$

When desired, as in the case of light hydrocarbons, the compressibility factor Z may be included in the formulas for gases or vapors as follows:

$$W = CKAP\sqrt{\frac{M}{ZT}}$$

Example 1

A safety valve bears a certified capacity rating of 3020 pounds of steam per hour for a pressure setting of 200 psi. What is the relieving capacity of that valve in terms of air at 100°F and 200 psi?

For steam

$W_s = 51.5KAP$

$3020 = 51.5KAP$

$KAP = 58.5$

For air

$W_a = CKAP\sqrt{\frac{M}{T}} = (356)(58.5)\sqrt{\frac{28.97}{460 + 100}}$

$W_a = 4750$ pounds per hour.

Example 2

It is required to relieve 5000 pounds of propane per hour from a pressure vessel through a safety valve set to relieve at a pressure of P_s pounds per square inch and with an inlet temperature of 125°F. What total capacity in pounds of steam per hour in safety valves must be supplied?

For propane (Since value of C is not known, use the conservative value $C = 315$.)

$$W = CKAP \sqrt{\frac{M}{T}}$$

$$5000 = 315 KAP \sqrt{\frac{44.09}{460 + 125}}$$

$$KAP = 57.7$$

For steam

$$W_s = 51.5 KAP = (51.5)(57.7)$$

$W_s = 2790$ pounds per hour set to relieve at P_s pounds per square inch.

Example 3

It is required to relieve 1000 pounds of ammonia per hour from a pressure vessel at 150°F. What is the total capacity in pounds of steam per hour at the same pressure setting?

For ammonia

$$W = CKAP \sqrt{\frac{M}{T}}$$

$k = 1.33$ ∴ from Table 6-1 $C = 350$

$$1000 = 350 KAP \sqrt{\frac{17.03}{460 + 150}}$$

$$KAP = 17.10$$

For steam

$$W_s = 51.5 KAP = (51.5)(17.10)$$

$$W_s = 880 \text{ pounds per hour.}$$

Pressure Relief Valve Specifications

In most cases pressure relief valve manufacturers prefer to verify the sizing of their products and/or will do the actual sizing and make recommendations. In order for the valve to be sized and a recommendation made, the following information must be supplied:

1. Materials of construction
2. Set pressure

3. Maximum inlet temperature
4. Allowable overpressure
5. Service — specify either the media to be handled or the following physical properties or both:
   A. Liquids — specific gravity (water = 1)
   B. Gases — specific gravity (air = 1)
   C. Vapors — molecular weight (if a mixture of vapors, the average molecular weight).
6. Backpressure — either constant at a specific psig or variable through a psig range
7. Required capacity
8. Code requirements, if any
9. Type of end connections
10. Preferred inlet size.

6.2 FLUSH BOTTOM TANK VALVES

General

Flush bottom tank valves are widely used for discharging materials from processing vessels. These valves are designed so that they completely seal off the outlet nozzle at the inner vessel wall and thus prevent solids from accumulating in the outlet nozzle neck, solidifying, and stopping discharge of the vessel contents. They also prevent the collection of unreacted material in the nozzle.

Construction of Valve

The valve bodies are a modified Y-type, with the stem being located in the vertical branch of the valve and the vessel contents discharged through the Y-branch. Seat and disc design is similar to that of a globe valve with metal-to-metal seating. There are two forms of this valve available: one in which the disc lifts up from the seat and opens into the vessel, and another in which the disc pulls down from the seat and opens into the valve.

The former style permits the breaking up of any encrustation which may form over the outlet opening. Some flush bottom valve manufacturers supply a second handwheel on the valve which is connected to the disc. This permits rotation of the disc as it is seated to wipe clean the seating surfaces as the valve is closed.

Consideration must be given to dimensions when selecting these valves. Flush bottom valves are sized by the *valve outlet size* which, in some cases, is *smaller* than the inlet connection which mates with the vessel nozzle. Most flush bottom valves require a specially sized tank adapter pad to permit installation of the valve. The valve manufacturer should be consulted for these dimensions. Notable exceptions to this are glassed-steel flush bottom valves which fit standard pipe size glassed-steel nozzles. If an alloy steel vessel is being fabricated for a specific installation, it is usually less expensive to design the outlet nozzle pad to fit the valve manufacturer's standards than it is to have the valve manufacturer supply a special valve.

Another style of valve consists of a piston or ram which extends up into the tank and thus prevents any possibility of the outlet becoming plugged. In the open position the piston is drawn down into the bonnet leaving a completely open passage for the material passing through. Since the piston occupies the entire internal area (less a few thousandths of an inch for clearance), closure insures a completely cleared valve ready for the next runoff. The close tolerance between the piston and surrounding parts leaves no room for condensation to accumulate and freeze when the valve is used on low temperature work. This valve has specific advantages in services where heavy sediments or crusts may deposit, particularly during the vessel emptying operation.

Stem Assemblies

The flush bottom valves having globe-valve type seat and disc configurations are supplied with outside screw rising stem handwheels. Ram- or piston-type flush bottom valves have an outside screw rising stem on the 1- and 1½-inch sizes but an outside screw nonrising stem on 2-inch sizes and larger. Position indicators are available.

Bonnet Construction

All of the flush bottom valves are fitted with stuffing boxes and packing to prevent stem leakage. In most instances TFE packing is supplied but other packing materials are available.

The piston-type valve has two resilient packing rings through which the piston moves. The lower ring prevents the flow from passing down into the bonnet, while the upper ring forms a seal as the piston enters it, thereby assuring a tight shutoff. To compensate for wear, these rings may be compressed by taking up on the compression nuts. When the rings are worn beyond the compression point, they should be replaced.

Identification of Parts

The nomenclature and identification of the parts and components of flush bottom tank valves are shown in Fig. 6-3.

Ch. 6 SPECIAL SERVICE VALVES 153

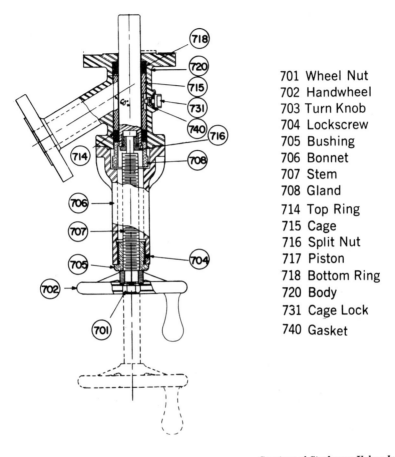

701 Wheel Nut
702 Handwheel
703 Turn Knob
704 Lockscrew
705 Bushing
706 Bonnet
707 Stem
708 Gland
714 Top Ring
715 Cage
716 Split Nut
717 Piston
718 Bottom Ring
720 Body
731 Cage Lock
740 Gasket

Courtesy of Strahmam Valves Inc.

Fig. 6-3. Ram-type drain valve.

Available Materials of Construction

Flush bottom tank valves are standard in:

Stainless Steel   Nickel
Hastelloy Alloys   Iron
Bronze   Carbon Steel
Monel   Aluminum

However, they can be furnished in practically any material of construction.

Installation and Operation

If a flush bottom valve is being installed which will open into the vessel, and the vessel is equipped with an agitator, care should be taken during the design stage to insure that the valve disc (or ram) will not contact the agitator. Consideration must also be given to the sizing of the bottom outlet nozzle and pad to insure a proper match between it and the valve.

Piston-type drain valves are available with air cylinder or electric motor operation on the larger size valves.

Drain valves in general are available from ½-inch to 6-inch sizes, in pressure ratings to 600 psig.

6.3 SAMPLING VALVES

Sampling valves are similar to flush bottom valves except for being considerably smaller. The ideal sampling valve seals at the inner wall of the pipe or vessel to which the valve is connected. When the valve is in the closed position, no material is present in the valve body or in piping leading to the valve. This insures that the sample being taken is representative of what is flowing through the pipe or is what is in the vessel at that moment.

These valves are available in most metallic materials of construction and usually range from ⅛ inch to 1 inch in size. Figure 6-4 shows a typical sampling valve in the closed and open positions.

6.4 SOLENOID VALVES

General

A solenoid valve is a combination of two basic functional units: an electromagnet or solenoid with its plunger or core, and a valve. The valve is opened or closed by movement of the plunger which is drawn into the solenoid when the coil is energized.

Valves are available in either a normally closed or a normally open construction. In the former the valve opens when current is applied (energized) and closed when the current is cut off (de-energized), while in the normally open construction the valve closes when current is applied and opens when the current is cut off. The terms normally open or normally closed refer to the valve position before current is applied.

Solenoid valves are designed for on-off operation and are either fully opened or fully closed.

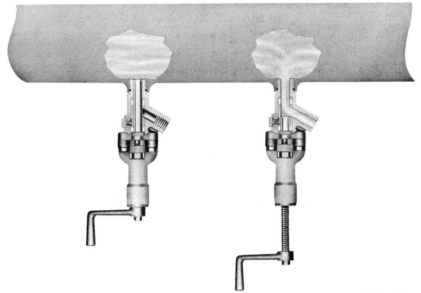

Courtesy of Strahmam Valves Inc.

Fig. 6-4. Typical sample valve installation.

Construction of Valve

Solenoids are usually employed with sliding stem on-off globe valves. Depending upon the application and service conditions, seating can be either metal-to-metal or composition disc to metal. There are four basic types of operation:

1. Direct acting
2. Internal pilot operated
3. External pilot operated
4. Semibalanced seat and disc.

Direct Acting Valve

In a direct acting valve the solenoid core (plunger) is mechanically connected to the valve disc and directly opens or closes the valve. A spring normally holds the plug in either the closed or open position, and it is against this force that the solenoid must move the plug to the opposite position. Operation is not dependent upon the line pressure or rate of flow.

Internal Pilot-Operated Valve

This valve is equipped with a small pilot orifice utilizing the line pressure for operation. When the solenoid is energized, it opens the pilot orifice and releases pressure from the top of the valve plug or diaphragm to the outlet side of the valve. This results in an unbalanced pressure across the plug or diaphragm which opens the main orifice.

When the solenoid is de-energized, the pilot orifice is closed and full line pressure is applied to the top of the disc providing a seating force for tight closure.

External Pilot-Operated Valve

Usually of diaphragm- or cylinder-operated construction, this valve is equipped with a three-way solenoid pilot which alternately applies pressure to or exhausts pressure from the diaphragm or cylinder for operation. Line pressure, or a separate source of pressure, such as air, can be used to operate the pilot valve.

Semibalanced Seat and Disc Valve

This is a form of a double seated valve. The body contains two seats, one above the other with a space between. The lower plug is slightly smaller than the upper plug. Both plugs are mounted on a single stem. Line pressure from the inlet side of the valve is introduced below the lower plug and above the upper plug. The downward force on the upper plug is greater than the upward force on the lower plug. The small differences in forces, plus a spring, hold the upper and lower plugs on their seats. When the solenoid is energized, the plugs are raised, opening the valve. Because of the force acting in an upward direction on the lower plug, the solenoid only has to overcome the slight differences in forces and the spring force.

Solenoid valves are also available in multiport configurations. Two-way valves are conventional line shut-off valves having one inlet and one outlet connection. The valve opens and closes, depending upon whether the solenoid is energized or de-energized.

Three-way solenoid valves have three pipe connections and two orifices. One orifice is always open and one orifice always closed. These valves are commonly used to alternately apply pressure to and exhaust pressure from a diaphragm valve or single acting cylinder, or they may be used to select or divert flow in pipe connections.

Four-way solenoid valves are generally used to operate double acting cylinders. These valves have four pipe connections — one pressure, two cylinder, and one exhaust. In one valve position, pressure is applied to one cylinder; the other is connected to exhaust. In the other valve position, pressure and exhaust are reversed.

Solenoid Enclosures

Enclosures for the solenoid coil are available to meet the various NEMA standards and requirements dependent upon the area of operation.

General Purpose NEMA Type 1

This type of enclosure usually consists of a pressed steel or cast-iron housing which has a clearance hole for standard conduit or BX fittings. The NEMA type 1 enclosure is suitable for general purpose applications indoors and where atmospheric conditions are normal. It serves as a protection against dust and light, indirect splashing, but is not dust-tight.

Watertight NEMA Type 4

These enclosures are suitable for applications outdoors, on ship docks, in dairies, breweries, and in effect wherever equipment will be subject to outdoor weather conditions.

Explosion-Proof Enclosures

This type solenoid enclosure is used in locations where flammable gases, vapors or dust are, or may be, present in sufficient quantities to produce explosive or ignitable mixtures. Depending upon the specific materials present, the enclosure must be listed for the appropriate hazardous location. Table 6-2 lists the classifications of hazardous locations in relation to the material present.

Table 6-2. Classification of Hazardous Locations

Classifi-cation	Class I		Class II		
	Group C	Group D	Group E	Group F	Group G
Material	Ethylether vapors Ethylene Cyclopropane	Gasoline Petroleum Naphtha Alcohols Lacquer Solvent vapors Natural gas	Metal dust	Carbon black Coal dust Coke dust	Grain dust

Stem Seals

Two basic types are available, the packless and packed valves. The packless valve has an enclosed solenoid mounted directly on the valve body with the solenoid core attached to the valve stem. The core is enclosed and free to move in a permanently sealed tube inside the solenoid coil. This construction provides a compact leak-tight assembly without the need of a stuffing box or sliding seal.

The packed-type valve has an enclosed solenoid or cylinder operator mounted on a bracket supported by the valve body with the valve stem operating through a rotary or sliding stuffing box. Generally a packed type valve is used wherever pressure, temperature, or nature of the material being handled prohibits the use of a packless valve.

Available Materials of Construction

Solenoid valves of the direct-acting type are available in most materials of construction including:

Cast iron	Stainless steel
Brass	Monel
Bronze	Alloy 20
Steel	Polyvinyl chloride

External pilot-operated valves are available in a much wider range of materials of construction since these are actually other types of valves such as diaphragm, ball, butterfly, etc., which are equipped with air operators or hydraulic operators. The solenoid pilot valve controls the flow of the air or hydraulic fluid to the valve operator, thus operating the valve.

Installation

Most solenoid valves may be mounted in any position without affecting operation. However, there are some solenoid valves which must be installed with the solenoid in the vertical upright position. The manufacturer's specific instructions should be checked before installation.

General installation techniques, described on page 40, should be followed for solenoid valves.

6.5 PRESSURE-REDUCING VALVES

Pressure-reducing valves are commonly used to reduce steam, gas, or liquid pressure to a predetermined and adjustable lower level. These self-contained units ordinarily consist of two valves built much like the diaphragm and spring-loaded relief valve.

A handwheel or adjusting screw compresses an adjusting spring against a metallic diaphragm opening a controlling valve. This valve admits high pressure from the inlet body port to the top of a piston or diaphragm, opening the main valve and admitting reduced pressure to the outlet piping. The reduced pressure acting through the outlet body port or through an external control pipe on the under side of the diaphragm balances the compression of the adjusting spring. This action throttles the controlling valve and controls the reduced or outlet pressure at the set value corresponding to the loading of the adjusting spring.

Any load change is accompanied by an immediate pressure change on the diaphragm, instantly repositioning the main valve to restore the reduced pressure to its set value.

Accuracy of Regulation

There is a definite relationship between accuracy of regulation and capacity of a reducing valve or regulator. A spring-loaded reducing valve should be adjusted while passing a minimum flow (not dead end). The reduced pressure obtained by slowly increasing the flow to valve rated capacity is a measure of accuracy of regulation. Therefore, a reducing valve set to deliver 100 psi pressure at minimum flow has a 99 percent accuracy of regulation if it delivers 99 psi with rated capacity.

Sensitivity

The sensitivity of a pressure-reducing valve is its ability to respond to pressure variations and to correct for load changes. It should not be confused with accuracy of regulation described above. To obtain the greatest sensitivity, pressure-reducing valves must be properly sized and selected for application, installed in accordance with manufacturer's instructions, and maintained so that internal parts move freely.

Valve Selection

The determination of the best valve for the job depends upon the specific application and service intended. Most manufacturers prefer to be consulted as to the best and most economical type to use. There are five major considerations which must be taken into account in sizing the valve. The manufacturer needs the answers to these questions.

1. *What are the minimum and maximum upstream pressures?*

The upstream pressure is the pressure that enters the valve, sometimes referred to as inlet or supply pressure. This upstream pressure should not be guessed at. If the valve is to be installed in an existing system, readings should be taken from a pressure gauge installed as close as possible to where the valve will be installed. If this is not possible, then the inlet pressure should be determined by calculating the pipeline losses between the supply source and the valve.

A common mistake made in sizing reducing valves for steam service is to use the boiler pressure and neglect the line losses. This can result in the selection of an undersized valve.

2. *What downstream pressure is to be maintained, or what adjustable range of reduced pressure is desired?*

Downstream pressure is the pressure discharged from the valve, sometimes referred to as outlet or reduced pressure.

The value of this pressure is determined by the process and usually involves no guess work. If the downstream pressure is fixed, the pressure differential established by the minimum upstream pressure can be used to size the valve. If an adjustable downstream pressure is required, then the valve must be sized according to the minimum pressure differential available.

3. *What minimum, maximum, and average flows of fluid are to be handled by the pressure-reducing valve?*

Do not buy a reducing valve to match the pipeline size. The pipeline size may or may not be the same as the proper size of valve to use. Each manufacturer has his own capacity tables to show the size to use for a given flow of fluid and design of valve.

Since designs vary greatly, capacities also vary for valves of the same pipe size. It is not unusual to find capacity tables showing considerable differences in capacities for a given pipe size, or differences in sizes for a given capacity.

In specifying the maximum flow there is a tendency to allow for an expected increase in load. It is unwise to allow more than 25 percent over actual requirement because oversizing the reducing valve may lead to difficulties in operating in the nearly closed position most of the time. Among these are unstable regulation, chattering, breakage of diaphragms, wire drawing of seats, and excessive wear.

The minimum flow to be handled is important. If a winter load of 90 percent heating steam and 10 percent process steam drops to a summer load of 10 percent process steam, it may be advantageous to use two reducing valves. Always check the minimum load to make sure that the valve is not oversized for this condition after having been chosen on the basis of the maximum load. Underloading a valve is synonymous with oversizing.

4. *Is dead-end shut-off required?*

A dead-end valve is one that closes tight, preventing flow of the fluid to the downstream side. This is a frequent requirement in main headers supplying steam to heating systems. If the valve does not close tight when steam is not required, it will permit the downstream pressure to build up to line pressure, or to a point where it will blow the downstream safety valve or damage the piping. A simple rule is: Do not use a double-seat reducing valve if a tight shut-off is required. Only single seat valves can supply a tight shutoff.

5. *What style of connection is required?*

This answer is determined by good piping practice and the actual conditions of installation. If the valve is to have threaded connections, it is recommended that pipe unions be installed on both sides of the valve.

Installation

The general rules for installation of valves found on page 40 also apply to pressure-reducing valves, as do the specific instructions which follow:

1. Good installation practice always includes a bypass to permit emergency repair or maintenance without shutting off the supply. Unions should be installed on both sides of valves having threaded connections.

2. Do not locate a reducing valve in an inaccessible spot. This makes it difficult to service and maintain.

3. Install a pressure gauge on both the inlet and outlet lines. These are required for setting and checking the reducing valve.

4. A certain amount of dirt and scale is always present in pipelines. Because of this it is advisable to install a Y-type strainer before the reducing valve.

5. Follow the manufacturer's specific instructions.

6. A safety valve should be installed downstream of the pressure-reducing valve.

Operation

1. When placing a reducing valve into service, check the position (open or closed) of all stop valves connected with the particular installation.

2. Open the strainer blow-off valve before cutting in the reducing valve to eliminate condensate and dirt that may have collected in the system. Condensate or slugs of water in a steam line are dangerous and may create shocks that will be harmful to working parts.

3. When placing a reducing valve into service, it is best to "crack" the downstream stop valve and then gradually open the upstream stop valve before setting the reducing valve. While this operation is being performed, watch the downstream pressure gauge for excessive pressure which might overheat or overstress the equipment. If excessive pressure becomes a problem, it can be readily controlled with the upstream stop valve.

4. Do not readjust or overset a reducing valve while filling the piping system. When a low-pressure system is cold, a reasonable time is required to bring it up to pressure, during which time the reducing valve will be wide open, until the desired pressure is reached.

If a reducing valve has been previously set, it should "take hold" and maintain the set pressure as the upstream stop valve is opened wide. The final adjustment can now be made on the reducing valve with the downstream stop valve still only cracked open. Then as this valve is opened wide, the pressure on the downstream gauge drops because of the heavy flow while the low-pressure system is being filled. When the pressure builds up to the previously set point, the valve throttles and maintains the pressure at a setting proportional to the flow at that time.

5. Before leaving a new installation, be sure that all stop valves are fully open and that all by-pass valves are tightly closed. A leaky valve in the by-pass line will give the impression that the reducing valve is leaking.

Maintenance

A properly installed and operated pressure-reducing valve will require little maintenance. Valves which have a stuffing box or packing gland should be inspected at least monthly for leaks. Glands should be kept tight to prevent leakage. In tightening packing glands be careful that excessive friction is not created on the stem so as to make the reducing valve sluggish or inoperative. If this condition develops, the stuffing box should be repacked. To test for a sluggish condition, close the downstream stop valve gradually and observe the movement of the stem, which should be free and not binding.

The most common solution for many problems, including an erratic or nonworking valve, is to take the valve apart, inspect it for wear, and clean it. Wear can occur in the seat, disc, and stems. The diaphragm may crack or become distorted. Springs may be corroded or broken. Replace any worn or damaged parts, clean thoroughly, and reinstall.

6.6 BACKPRESSURE REGULATING VALVE

General

Backpressure regulating valves are self-contained devices which prevent pressure in a system from exceeding a predetermined level by relieving into a low-pressure line. These regulators are very similar in design to relief valves but contain design refinements that give better control in continuous or intermittent service. Relief valves are designed for emergency service only and are not suitable for continuous service.

Construction of Valve

Backpressure regulating valves have a spring-loaded disc or piston that lifts to open the valve when the pressure in the system rises above that exerted by the spring. The pressure is not usually applied directly to the disc or piston but rather to a diaphragm that acts on the spring, which moves the disc stem to open or close the valve. The desired control pressure is set by compression or decompression of the spring via a screw at the top of the valve.

6.7 CRYOGENIC VALVES

General

The use of liquified gases (oxygen, nitrogen, hydrogen, helium, etc.) at cryogenic temperatures has prompted the development of valves to handle these materials safely and economically. Valves for this service range in size from ⅛ inch to 30 inches, with pressure ranges from full vacuum to 10,000 psig and temperatures as low as minus 455°F.

Valves for this service are usually of bronze or 300 series stainless steel construction of the gate, globe, needle, butterfly, or check design. They differ from the conventional design of the above valves in that they all have extended bonnets. They are so designed that there is no possibility of the stuffing box freezing and causing the valve to become inoperable. This extension also minimizes heat leakage through the stem and provides insulation space between the pipeline and the handwheel.

Where extremely tight shutoff is required, the discs are fitted with TFE or Kel-F[1] inserts. Valve stem packing is also usually TFE. On valves where heat loss presents a problem the valve should be equipped with vacuum jackets.

6.8 CONTROL VALVES

The subject of automatic control valves is a broad and distinct field in itself — separate from that of manually operated or power-operated but manually controlled valves. It is not intended that coverage be given to these valves in this book. In general, any of the valves discussed in this book which can be fitted with automatic operators can be incorporated into an automatic control circuit with the valve operation being performed on signal from a control unit which may be controlling temperature, pressure, flow, liquid level, or some other process variable. Usually control valves are made of the single or double plug globe-type, needle-type, diaphragm-type or butterfly-type valves.

[1] Trademark of 3-M Company.

7

Valve Packing

The majority of the valves in service utilize stuffing boxes to produce the stem seal. In order for the stem seal to be effective it is essential that the correct type of packing and that the correct packing material be used for the conditions of service.

When selecting packing materials it is necessary to take into account the following factors:

1. Material being handled
2. Operating pressure
3. Operating temperature
4. Minimum temperature (For example a valve may be handling steam at 300°F but during off periods the line may be at ambient temperature.)
5. Size of packing
6. Materials of construction of the valve stem.

Care should be used in the selection of packings for use with stainless steel valve stems. Stainless steel has solved many mechanical and chemical problems, but graphite uninhibited, used to lubricate many packings, is not suitable for use with stainless steel because it "pits" the metal. To overcome this difficulty either mica, or a sacrificial metal, or TFE is used as the packing lubricant.

Proper installation of packings is of equal importance to proper selection of the packing material and style. Extreme care should be taken to insure that the packing to be installed is free of air contamination and foreign matter. Manufacturers take precautions during the manufacturing operations to prevent any contaminating elements from getting into the packing. Each package is usually carefully wrapped so that when unpacked for use, the only requirement is that the stuffing box be free of old packing material, dirt, and corrosion.

Braided and Twisted Packings

Braided and twisted packings are produced from a variety of materials, including long fiber spinning grades of white asbestos, African blue asbestos, and a variety of vegetable and synthetic fibers. Each of the packings produced is lubricated with various types of lubricants depending upon the specific application.

In general these packings are available in four styles:

1. *Regular Braid* (also called *Square, Plaited,* or *Flax Braid*). Each strand passes over and under strands running in the opposite direction. Cross sections are square. These packings are especially suited for high-pressure valve-stem packing and high-speed rotary and reciprocating service.

2. *Braid-over-Braid* (also known as *Multiple* or *Round Braid*). This packing is built up to required size by braiding one or more covers around a central core of braided, twisted, or homogenous materials. Especially good for valve stems where out-of-square cross sections are desired.

3. *Twisted Packings.* Yarns are twisted around each other to obtain the desired size. One packing size can be used for stuffing boxes of various sizes. Because strands are easily undone, they may be twisted to the desired size by using only the number of strands needed. This is an especially good general-utility or emergency-type packing where packing space is small.

4. *Diagonally Braided Packing* (also found under manufacturers' trade names such as *Lattice Braid,*[1] or *Chemlock*[2]). Diagonally braided packing is braided inside as well as outside. Each strand passes diagonally through the body of the packing at an angle of approximately 45 degrees. This produces a completely unified structure. Every braided strand contributes to the strength of the entire packing.

Diagonal braiding makes each strand much more flexible than ordinary braiding. Thus there is less stress when it is formed into rings. The strands are individually lubricated to distribute lubricant evenly throughout the packing and to control proper porosity. Because of its increased flexibility, this style of packing is ideal for application around small-diameter valve stems.

---

[1] Registered trademark of Garlock Inc.
[2] Registered trademark of Chemical & Power Products Inc.

Each style of packing is available in die molded rings or spiral form. The die molded rings are preformed and cut to specific inside diameters with various packing thicknesses. The spiral form is available in various lengths depending upon the specific material and the manufacturer.

Fabric and Rubber Packings

Fabric and rubber packings are produced from a variety of different weight cotton duck fabrics and asbestos with natural, neoprene, or SBR rubber binders. Construction features of the cotton duck and rubber packing vary somewhat from the features of the asbestos and rubber packings. All types are available in die-molded rings, spiral and coil forms.

The duck and rubber packing is usually available in three styles:

1. *Laminated Construction,* where the duck and rubber are press-laminated into slab form, then fabricated into either coil, spiral, or ring configuration. The material is laminated either flat or diagonally.

2. *Rolled Construction* from a medium-weight rubber-coated cotton duck with a hollow center. The cross section is square. This permits expansion and contraction of packing without binding.

3. *Rolled Construction* from a medium-weight rubber-coated cotton duck rolled around a round rubber core, producing a round cross section. High resiliency permits flexing without binding or seizing.

The asbestos and rubber packings are usually available in two styles:

1. *Laminated Construction,* where the asbestos and rubber are laminated into slab form, then made into coils, spirals, or rings.

2. *Rolled Construction,* made of rolled asbestos cloth and rubber or metal in three styles: with rubber core, with lead wedge, or with accordion-folded asbestos cloth and rubber strip.

Metallic Packings

Metallic packings are available in a semimetallic and metallic or foil constructions. Semimetallic packings are produced by wrapping metallic foil over a nonmetallic core such as asbestos. Metallic or foil packings are produced by twisting and wrapping specially treated lubricated crinkled sheets upon themselves. Metallic packings should not be used on brass valve stems or any valve stem where the base metal of the valve stem and packing are of a similar material. These packings are recommended for high temperature and high pressure resistance. They are available in spiral or die-molded ring forms.

Plastic Packings

One form of plastic packing consists of compounded packings usually made of asbestos fibers combined with binding materials and lubricants. Some styles contain soft bearing metals. This mixture is then extruded in a rectangular cross-section, in spiral form. The packing thus produced is soft, readily formable, and particularly suited to sealing gases and highly mobile liquids.

Two other types of plastic packing, which were developed primarily for use with highly corrosive materials, are various TFE packings and TFE impregnated packings.

TFE packings are produced from a continuous TFE filament in the same style as conventional braided and twisted packings. Some styles, after braiding, are impregnated with a TFE suspensoid to produce a 100 percent TFE packing. Some packing is also produced by extrusion of the TFE resin. Depending upon application, additives may be used in the extrusion operation.

TFE-impregnated packings consist of asbestos or Blue African asbestos yarn impregnated with TFE. In some styles each individual strand of asbestos is impregnated with TFE prior to braiding — other styles are produced by impregnating an already formed braided packing. For extremely critical services each strand is impregnated, and after braiding reimpregnated with TFE suspensoid.

Die-Molded Packings

Also available are the same wide choice of previously described packing materials but in a molded form. Molded packings are formed shapes of packing materials which by virtue of their shape and size rely on fluid pressure to effect a seal. This tends to eliminate periodic gland adjustment, with only the initial gland adjustment being required. The advantages of molded packings versus "jam type" packings are minimum leakage with relatively low friction and long service life.

There are two general types of molded packings:

1. Squeeze-type
2. Lip-type.

"O"-rings, T-shaped rings, and other similar configurations that depend upon radial compression (dimensional interference) for the initial low pressure seal and on the fluid pressure for the high pressure seal are of the squeeze-type. "V"-rings, cups and flange packings, and other similar forms, the lips of which are held against the valve stem by fluid pressure, are of the lip-type. Low pressure sealing is usually accomplished by means of a slight flare molded into the lip which provides an initial interference. As the fluid pressure increases it provides additional force to the flare to effect a seal at the higher pressures.

Molded packings are precision products and as such are usually designed for specific applications. Stuffing box and stem dimensions must be carefully matched against the dimension for which the molded packing has been designed to operate.

Installation of Packing

After the packing which meets the operating conditions has been selected, there is still one important factor to be considered — proper installation. Overtightening of glands, using undersized or oversized cross sections and seating rings improperly are just a few of the possible causes of packing failure. The following general procedures and specific procedures for installation of either die-molded or spiral packing should be adhered to in order to get maximum life and efficiency from the packing.

General Procedures

1. Always store packings in a clean, dry place. Dirt and abrasive materials should never come in contact with packing before installation.
2. Use all new packing. Never install used rings.
3. Prior to installing new packing, clean stuffing box thoroughly.

Installation of Die-Molded Ring Sets

1. Check to see that packing is beveled if stuffing box is beveled and square when stuffing box is square.
2. See that the stem is not scored.
3. Split the first ring around the valve stem and force in firmly and evenly to the bottom of the box. Use a split cylinder or a tamping tool for this.
4. Be sure to insert rings in the order assembled with the joints staggered at 90 or 120 degrees. Compress each ring firmly as above.
5. With box full, compress with the gland, then back off the nuts half a turn. Be sure to adjust the gland evenly if the gland is not of the single packing gland nut variety.

Installation of Spiral Packing

1. Cut packing on mandrel or shaft by winding packing needed for stuffing box around rod and cut through each turn while coiled. The cuts should be diagonal. Packings should be cut to fit exactly.
2. See that the valve stem is not scored.
3. Split the first ring around the valve stem and force it firmly and

evenly to the bottom of the box. Use a split cylinder or a tamping tool for this.

4. Apply the remaining rings with the joints staggered at 90 or 120 degrees. Compress each ring firmly as described above.

5. When the box is full, compress the packing with the gland, then back off the nuts half a turn. Be sure to adjust the gland evenly if the gland is not of the single packing nut variety.

6. After packing the stuffing box, wrap and replace all unused packing in container as a protection against waste and picking up of dirt or foreign substances.

Packing Recommendations

Table 7-1 provides a listing of the more common materials handled and a series of compatible packings. It must be recognized that this listing is only a guide since valve construction and actual conditions of operation will affect the ultimate packing choice. However, it does provide a source from which the most probable suitable materials may be selected. For extremely critical and/or hazardous applications it would be advisable to verify the selection with the packing manufacturer.

Table 7-1. Packing Recommendations*

Style-of-Packing Key
A = Diagonally Braided Packing   E = Fabric and Rubber Packing
B = Braid over Braid   F = Metallic Packing
C = Regular Braid   G = Plastic Packing
D = Twisted Packing

Service	Recommended Packing Style						
	A	B	C	D	E	F	G
Acetic Acid, Glacial	8, 6	12, 14				38	36
Acetic Acid, Dilute	8, 6	12, 13				38	29, 30
Air (Max. 500°F)	1, 2, 6, 7	11	18		22, 41	24, 26, 38	
Air (Max. 700°F)	2, 6, 7	11				26, 38	
Alcohols	6, 8	12, 13				26, 38	35
Aliphatic Solvents (Gasoline, etc.)	6, 8	12					35, 36
Alum Liquor	8					24, 37	
Ammonia, Aqueous	8	12, 13			21	24, 26, 37	
Ammonia, Gaseous	8					24, 26, 37	

Table 7-1 (continued). Packing Recommendations*

	Recommended Packing						
	Style						
Service	A	B	C	D	E	F	G
Ammonia, Liquid	8	12, 13				24, 26 37	
Aromatic Solvents (Benzene Chlorinated Types, Carbon Tetrachloride, etc.)	6, 8	12, 15	16				34, 36
Brine	8		20	39			
Carbon Bisulphide	6, 8	15	16				
Carbon Dioxide				39	21		
Caustics	4, 7, 8						31
Chlorine	6, 8	12			23		
Crude Oil						24, 26, 37, 38	
Dowtherm A						26, 38	
Dowtherm E						27, 28	
Esters	6, 8	15	16				35, 36
Fuel Oil, Heavy						24, 37	
Gas Illuminating				39	41		
Gasoline	3, 6, 8	15	16				
Halogens	6, 8	12					
Hydrachloric Acid Conc.	6, 8	12, 14					31, 34 35, 36
Hydrachloric Acid Dilute	6, 8	12, 13					35, 36
Hydrogen					41		
Inert Gases (Max. 500°F)	1, 6, 7		18				
Ketones	6, 8	12, 13					35, 36
Lubricating Oil						24, 37	
Mineral Acids	4, 6, 7, 8						
Nitric Acid	6, 8	12					31, 34, 36
Oil, Cold	3	15, 40	16			24, 26	32, 33
Oil, Hot (700°F Max.)	3	15, 40				24, 26	32, 33
Oil, Hydraulic (Max. 550°F)						25	

Table 7-1 (continued). Packing Recommendations*

Service	Recommended Packing Style						
	A	B	C	D	E	F	G
Organic Acids, Dilute	6, 8	12, 13					
Oxygen, Gaseous	5, 9	10					
Oxygen, Liquid	5, 9						
Phosphoric Acid, Pure	8					24, 37	
Soda Ash	8					24, 37	
Steam, Low Pressure	13, 17	11, 19			21, 22	24, 27	29, 32
Steam, Med. Pressure	13, 17	11, 19		39	22, 41	24, 27, 37	29, 32
Steam (Max. 500°F)	1, 6, 7	11, 19			41	24	29, 30, 32
Steam (Max. 1200°F)		11, 19					
Sulfite Liquor	8					24, 37	
Sulfuric Acid, Conc.	8	12, 14				37	31, 34, 35, 36
Sulfuric Acid, Dilute	8	12, 13, 14				37	31, 34, 35, 36
Sulfurous Acid	8	12, 13				37	31, 34, 35, 36
Water, Cold		13, 17	18, 20	39	21	24, 25, 27, 37	30
Water, Hot		11, 13, 17, 19	18	39	21	24, 25, 27, 38	30

* A description of each packing number will be found in Table 7-2.

Table 7-1 lists the recommendations by number. A description of the specific packing numbers will be found in Table 7-2. Since each packing manufacturer produces his packing with a slightly different formulation, the packing descriptions are somewhat general.

Table 7-2. Packing Description

Packing Number*	Description	Max. Oper. Temp.
1.	Long fiber asbestos yarn, graphited throughout	500°F
2.	Long fiber asbestos yarn with copper wire insertion and copper wire corners, graphited throughout	700°F
3.	Long fiber asbestos yarn lubricated with gasoline-resistant compound and graphited throughout	700°F

Table 7-2 (continued). Packing Description

Packing Number*	Description	Max. Oper. Temp.
4.	Blue African asbestos yarn graphited throughout	
5.	Blue African asbestos yarn with TFE impregnation, specially treated	450°F
6.	Long fiber blue African asbestos yarn with TFE impregnation	450°F
7.	Long fiber asbestos yarn with TFE impregnation	450°F
8.	TFE fiber yarn with TFE impregnation	450°F
9.	TFE fiber yarn with TFE impregnation, specially treated	450°F
10.	Pure white asbestos yarn neither lubricated nor graphited	
11.	Core of white asbestos, inconel wire inserted. Surface treated with an elastomeric material and graphite	1200°F
12.	Blue African asbestos impregnated with TFE, no graphite	
13.	Long fiber white asbestos yarn impregnated with mineral-fat lubricant; graphited throughout	
14.	Blue asbestos treated with mineral animal-fat lubricant, externally graphited	
15.	Long fiber white asbestos yarn with soap-fat lubricant; graphited throughout	
16.	Long fiber white asbestos yarn lubricated with soap and glycerol compound and graphited throughout	
17.	Long fiber white asbestos yarn with mineral-fat lubricant; graphited throughout	
18.	Long fiber white asbestos yarn, copper wire inserted, impregnated throughout with a petroleum lubricant; graphite-coated surface	
19.	Long fiber white asbestos yarn. No internal lubricant but graphite-coated on the surface	750°F
20.	Jute fiber treated with mineral oil, paraffin, and marine-fat lubricant; ungraphited	
21.	Laminate of heavy-weight cotton duck treated with SBR rubber compound. Graphited and mineral oil lubricated	
22.	Fabric jacket around a core of accordion-folded rubber-impregnated cloth. Externally treated with glycerol lubricant and graphite	600°F

Table 7-2 (continued). Packing Description

Packing Number*	Description	Max. Oper. Temp.
23.	Asbestos cloth frictioned with SBR rubber, treated with beeswax lubricant. Surface graphite coated	600°F
24.	Lead foil twisted around an asbestos core, lubricated and graphited	500°F
25.	Lead foil spiral-wrapped around center core of braided and lubricated asbestos	550°F
26.	Lubricated and graphited aluminum foil twisted around an asbestos core	1000°F
27.	Lubricated and graphited copper foil twisted around an asbestos core	1500°F
28.	Copper tinsel bonded with rubber cement in regular braid; surface coated with graphite	1500°F
29.	Asbestos fibers and graphite, bonded with a water-resistant, oil-resistant binder with corrosive inhibiting agent	600°F
30.	Asbestos fibers, lead particles, and graphite bonded with a water- and oil-resistant binder with a corrosion inhibiting agent	600°F
31.	Blue African asbestos fiber with a bonding agent containing acid-resistant lubricant and graphite	
32.	Asbestos fiber with a bonding agent containing lubricant and graphite	600°F
33.	Asbestos fiber, shredded copper foil, and bonding agent containing lubricant and graphite	600°F
34.	Plastic packing of TFE powder, graphite, and synthetic hydrocarbon binder, with skeleton braided covers of TFE yarn	500°F
35.	Dry packing of pure TFE. No binder	450°F
36.	Shredded TFE impregnated with TFE suspensoid	500°F
37.	All lead foil twisted and lubricated	500°F
38.	All aluminum foil twisted and lubricated	1000°F
39.	Strands of long fiber asbestos yarn impregnated with high-temperature petroleum lubricant and flake graphite	
40.	Long fiber asbestos yarn with monel wire insertion braided over a high temperature plastic core	
41.	Core of asbestos cloth and rubber cushion with outer cover of heat resistant rubber-frictioned asbestos cloth lubricated and graphited	600°F

* Numbers as given in Table 7-1.

Appendix

Recommended Valve Services

Valve	On-Off	Throttling	Diverting Flow	Freq. Oper.	Low Press. Drop	Slurry Handling	Quick Opening	Free** Draining	Prevent Reversal of Flow	Prevent Over-pressure	Control Pressure
Gate	X				X		X	X			
Globe	X	X*		X							
Plug	X	X*	X	X	X		X	X			
Ball	X	X*	X	X	X		X				
Butterfly	X	X*		X	X	X	X	X*			
Diaphragm	X	X*				X	X*	X*			
Y	X	X*		X							
Needle		X									
Pinch	X	X			X	X		X			
Slide	X				X	X	X*	X			
Swing Check									X		
Tilting Disc Check									X		
Lift Check									X		
Piston Check					X				X		
Butterfly Check									X		
Spring Loaded Check									X		
Foot Valve									X		
Stop Check		X									
Pressure Relief	X									X	
Pressure Reducing		X									X
Back Pressure										X	X
Sampling	X										

* Certain configurations only. Check detailed section.
** All of these valves may not be completely free draining, but they trap a minimum amount of fluid.

Index

Abbreviations, valve, 7–12
Air motors, 34
Air operators, 34
Angle valves, 66
ASTM designations of valve metals, 5–7

Back pressure regulating valve, 162
   construction, 163
Ball valve, 87–93
   construction, 87
   identification of parts, 90
   installation, 40, 91
   maintenance, 91
   materials of construction, 90
   multiport, 88
   operation, 91
   seats, 89
   service recommendations, 57, 87
   specifications, 93
   stem and bonnet, 89
Bellows seal, 21
Bolted bonnet, 24
Bonnet, bolted, 24
   breech lock, 27
   flanged, 24
   lip sealed, 26
   pressure sealed, 26
   screwed, 22
   seals, 22
   U-bolt, 25
   union, 23
Braided packing, 165
Braid-over braid, 165
Butterfly check valve, 129–132
   construction, 130

Butterfly check valve, installation, 117, 132
   maintenance, 132
   materials of construction, 131
   service recommendations, 130
Bypass connection, 45

Care of valves, 40
Check valve, 116–135
   installation, 40, 117
   maintenance, 117
   operating ranges, 117
   size ranges, 117
   types, 116
Coefficient, flow, 50
   resistance, 49
Connection, bypass, 45
   drain, 45
Control valves, 163
Critical applications, 10
Cryogenic valves, 163

Darcy's formula, 48, 53
Diagonally braided packing, 165
Diaphragm seal, 21
Diaphragm valve, 102–112
   body materials, 104
   construction, 103
   diaphragm materials, 106
   identification of parts, 108
   installation, 40, 108
   maintenance, 108
   operation, 108
   service recommendations, 59, 103
   specifications, 112
   stem and bonnet, 106

INDEX

Diaphragm valve, vacuum service, 107
Die molded packing, 167
Disc, butterfly valve, 94
   composition, 67
   double, 61
   metal, 68
   plug type, 68
Drain connection, 45

Electric motor operators, 36
Electric solenoid operators, 36
Equivalent lengths (Table), 50
Extension stem, 29

Fabric and rubber packing, 166
Flexible wedge, 60
Floor stands, 30
Flow coefficient, 50
Flow control elements, 13
Flow, laminar, 47
   turbulent, 47
   viscous, 47
Flow seals, 18
Flush bottom valve, 151–154
   bonnet, 152
   construction, 152
   identification of parts, 152
   installation, 154
   materials of construction, 153
   operation, 154
   ram type, 152
   stem assembly, 152
Foot valves, 134
Full flow, 9
Functions of valves, 1

Gate valve, 58–66
   bonnet, 22, 62
   construction, 59
   identification of parts, 63
   installation, 40, 65
   maintenance, 65
   materials of construction, 65
   operation, 65
   seat, 62
   service recommendations, 57–59
   specifications, 66
   stem assembly, 62
   stem seal, 62
Gear operators, 32
General purpose valves, 57
Globe valve, 66–71
   bonnet, 22, 70
   composition disc, 11, 67
   construction, 67
   identification of parts, 71

Globe valve, installation, 40, 71
   maintenance, 71
   materials of construction, 71
   metal disc, 68
   operation, 71
   plug-type disc, 65–70
   seat, 70
   service recommendation, 57, 67
   specifications, 76
   stem assembly, 70
   stem seal, 70

Hammer-blow handwheel, 32
Hydraulic operators, 34

Inside screw non-rising stem, 15
Inside screw rising stem, 15
Installation, angle valve, 40, 71
   ball valve, 40, 91
   butterfly valve, 40, 99
   butt weld end valve, 43
   check valve, 117
   diaphragm valve, 40, 108
   flanged end valve, 42
   gate valve, 40, 65
   globe valve, 40, 71
   needle valve, 40, 71
   pinch valve, 40
   plug valve, 40, 85
   slide valve, 40
   socket weld end valve, 42
   threaded end valve, 41
   valve, 40
   Y-type valve, 40, 71
Installation position, 11

Laminar flow, 47
Length equivalent (Table), 50
Lift check valve, 125–128
   construction, 125
   identification of parts, 127
   installation, 117, 128
   maintenance, 128
   materials of construction, 128
   service recommendations, 125
Lip sealed bonnet, 26
Location of valves, 39

Metallic packings, 166
Metal-to-metal seats, 11
Motors, air, 34

Noncritical applications, 10

Operating ranges of valves, 58
Operators, air, 34

INDEX 179

Operators, gear, 32
  hydraulic, 34
  wheel, 32
Outside screw rising stem, 16

Packing, braided, 165
  die molded, 167
  fabric and rubber, 166
  installation, 168
  metallic, 166
  plastic, 167
  recommendations, 169
  twisted, 165
Packing box seal, 18
Packing of valves, 164–173
Pinch valve, 112
  construction, 113
  identification of parts, 114
  installation, 40
  materials of construction, 114
  service recommendations, 57, 113
  specifications, 114
Piston check valve, 129
Plastic packings, 167
Plug valve, construction, 78
  gland design, 82
  identification of parts, 82
  installation, 40, 85
  lubricated, 80
  lubrication, 85
  maintenance, 85
  materials of construction, 85
  multiport, 78
  non-lubricated, 82
  operation, 85
  plug design, 78
  service recommendations, 57, 77
  specifications, 86
Position indicators, 33
Pressure drop, 48, 53
Pressure reducing valve, 158–162
  installation, 40, 161
  maintenance, 162
  operation, 161
  selection, 159
Pressure relief valve, 136–151
  accumulation, 138
  ASME-UPV formula, 139
  backpressure, 139
  blowdown, 138, 142
  capacity conversions, 147
  cold differential test pressure, 138
  construction, 142
  differential set pressure, 138
  installation, 145
  lift, 139

Pressure relief valve, maintenance, 145
  materials of construction, 145
  maximum allowable working pressure, 138
  operating pressure, 138
  orifice area, 139
  overpressure, 138
  reactive force, 141
  set pressure, 137
  sizing, 139
  specifications, 150
Pressure sealed bonnet, 26

Reactive force, 141
Regular braid, 165
Relief valves (see *Pressure relief valve*)
Replaceable disc, 10
Replaceable seat, 11
Resistance coefficient, 49
Reynolds number, 47, 52

Safety relief valves (see *Pressure relief valve*)
Safety valves (see *Pressure relief valve*)
Sampling valves, 154
Screwed bonnet, 22
Seal, bellows, 21
  bonnet, 22
  diaphragm, 21
  flow, 18
  packing box, 18
  stem, 21
  stuffing box, 18
Seat, ball valve, 89
  butterfly valve, 94
  gate valve, 62
  globe valve, 10, 11, 70
Service recommendations, angle valve, 57, 67
  ball valve, 57, 87
  butterfly valve, 57, 93
  check valve, 117
  diaphragm valve, 57, 103
  gate valve, 57–59
  globe valve, 57, 67
  needle valve, 57, 67
  pinch valve, 57, 113
  plug valve, 57, 77
  slide valve, 57, 115
  Y-type valve, 57, 76
Size ranges of valves, 58
Slide valve, 115
  service recommendations, 57, 115
Solenoid valve, 154–158
  construction, 155
  direct acting, 155

INDEX

Solenoid valve, enclosures, 157
  NEMA Type 1, 157
  NEMA Type 4, 157
  explosion proof types, 157
  external pilot operated, 156
  installation, 40, 158
  internal pilot operated, 155
  materials of construction, 158
  multiport, 156
  semi-balanced seat and disc, 156
  stem seals, 157
Solid wedge, 59
Sonic velocity, 51
Specifications, valve, 4
Split wedge, 60
Spring-loaded check valve, 132–134
  construction, 133
  installation, 117, 134
  maintenance, 134
  materials of construction, 133
  service recommendations, 133
Standards, valve, 4, 5
Stands, floor, 30
Stem, extension, 29
  seal, 18
Stop check valves, 134
Straight-through flow, 9
Stuffing box seal, 18
Swing check valve, 119
  construction, 119
  identification of parts, 121
  installation, 117, 122
  maintenance, 122
  materials of construction, 121
  service recommendations, 119

Terms, valve, 9
Throttled flow, 9

Tilting disc check valve, 122–125
  construction, 123
  identification of parts, 124
  installation, 117, 124, 125
  maintenance, 117
  materials of construction, 124
  service recommendations, 123
Trim, valve, 9
Turbulent flow, 47
Twisted packing, 165

U-bolt bonnet, 25
Union bonnet, 23, 24

Valve, abbreviations, 7
  angle, 66
  butt weld end, 38
  connections, 36
  flanged, 37
  functions, 1
  general purpose, 57
  silver brazing end, 38
  socket weld, 38
  solder end, 38
  specifications, 4
  standards, 4
  terms, 9
  threaded, 37
  trim, 9
Viscosity equivalents, 53
Viscous flow, 47

Wedge, flexible, 60
  solid, 59
  split, 60
Wheel operators, 32
Wire drawing, 10